Welding and cutting

Welding and cutting

A guide to
fusion welding and associated cutting processes

Peter Houldcroft, FEng, FIM, FWeldI
and
Robert John, PhD, CEng, FIM, FWeldI

Published in association with
The ESAB Group

Industrial Press Inc
New York

Published by Industrial Press Inc.,
200 Madison Avenue, New York,
New York 10016, USA

First published 1989

Library of Congress Cataloging in Publication Data
Houldcroft, P. T. (Peter Thomas)
Welding and cutting: a guide to fusion welding and associated cutting processes / by Peter Houldcroft, Robert John.
p. cm.
Bibliography: p.
Includes index.
ISBN 0-8311-1184-4
1. Welding—Handbooks, manuals, etc. I. John, Robert, 1948–
II. Title.
TS227.H676 1989
671.5′2—dc19 88-24084
CIP

Printed in Great Britain

ISBN 0-8311-1184-4

Contents

Preface

The ability to weld and cut metals lies at the heart of many of the industrial activities that now serve our way of life. The availability of such essentials as energy, food, transport chemicals, etc. is to a large degree dependent on structures and equipment that are created by the application of welding and cutting technology. This technology can be complex and draws on expertise from many different disciplines.

Basically, metallurgical 'know-how' is demanded for the selection and design of consumable filler materials and finished weld metals, whereas the design of the equipments used to create electric arcs or other heat sources is mainly through electronic and electrical engineering. These technologies are entirely different but must be brought together in the welding process itself. Other technologies and disciplines, for example, mechanical engineering, data processing and physics, can also have an impact on the design and selection of welding processes.

Knowledge of applying welding together with overall expertise in the field is often covered by the term 'welding engineering'. The scope of this book which focuses on fusion welding and associated cutting processes therefore covers a number of specialist topics.

The purpose of the book is to provide a structured reference work for those who are associated with the selection and application of materials and equipment for welding and cutting. We have had to strike a balance between describing first principles which can be absorbed by a general target audience who have little familiarity with the subject, and more complex technical descriptions which are suited to those in need of more detailed information. For those who require further detail we have included a note on sources of information and a list of reference works. It is envisaged that the contents would be particularly useful for those familiar with one field or aspect who require a general working knowledge in another.

Particular gratitude must be expressed to ESAB AB, ESAB Group (UK) Ltd, who have sponsored the book, and to colleagues from this organisation who have willingly given their time and expertise in reviewing the contents, and to The Welding Institute who have provided information and a number of illustrations. Also thanks are due to the Welding Institute's journal *Metal Construction*, for the use of illustrations.

P. T. Houldcroft
R. John

Chapter 1

Background and scope

Historical development

Until the late nineteenth century the only welding process was that used by the blacksmith. The smith was able to some extent to concentrate the heat where he wished to make a joint, by heaping the heated coals around this part. But it was not until the development of the electric arc and high temperature gas blowpipes that local heating and melting could be achieved and welding, as we know it now, became possible. Modern welding is a technology based on heat sources, of which there are many capable of being used to raise metal locally to a temperature suitable for welding. It is a highly efficient, economical and adaptable method of joining which has given designers and engineers new freedom in the use of materials. Nowhere is this more noticeable than in the elegant appearance and clean lines of large steel structures (see Fig. 1.1).

The first really practical welding process used the heat of an arc between a carbon electrode and the workpiece. This was the subject of UK patent No. 12984 of 1885 in the names of Benardos and Olszewski. The heat of the arc melted the metal where the joint was required and when the arc was removed the heat flowed away into the surrounding metal causing the molten pool to solidify, thereby unifying the pieces.

Some years later arc welding was developed further when a steel rod was substituted for the carbon electrode and now the process of heating was accompanied by the deposition on to the workpiece of metal melted from the end of the rod.

With the availability of oxygen and acetylene high temperature flames were made possible which could also melt metal locally. The intensity of all these heat sources enabled heat to be generated in the workpieces faster than it was being conducted away into the cold surrounding metal. A molten pool could be generated which when allowed to solidify formed the unifying bridge between the parts to be joined.

Fig. 1.1 *An elegant modern welded structure – the Wye approach bridge to the Severn bridge.*

Other electric welding processes which made use of the higher local electrical resistance where two pieces of metal were pressed together were also developed over the same period of time. These welding processes go under the general title of 'electric resistance welding' and although melting does in fact take place in some of them they are not classed as fusion welding processes, and will not be discussed here. The term 'fusion welding' is reserved for those processes where a heat source causes local melting (as a result of heat being added faster than it is conducted away) and the resultant molten pool forms a bridge between the parts when it solidifies. Pressure between the parts such as is employed in the resistance welding processes is not used at all in fusion welding.

Experience with the early arc processes using either a carbon electrode or a bare iron wire quickly indicated that if the arc was too long the fused metal had a tendency to be brittle. This was thought to be a result of the reaction of the hot metal with oxygen and particularly nitrogen in the atmosphere. Too much emphasis should not be put on this, however, as the reason for the initial use of coatings on electrode wires was almost certainly to improve the performance of the arc from the electrical point of view. The fact that heavier coatings on steel wires confer protection from the atmosphere was an advantage appreciated later and both the simple carbon-arc and bare-wire metal-arc processes remained in use up to the late 1930s.

An important advance was made in 1909 when Oscar Kjellberg first coated bare steel welding electrodes. He formed a company which very successfully exploited the development and which is now known internationally as ESAB. In Britain this Swedish development stimulated, by a fascinating series of events, the use of blue asbestos coverings for electrodes by a chemist, Arthur Stromenger, who formed the Quasi-Arc company. The asbestos wound electrode was easy to use and the Quasi-Arc company was instrumental in advancing to a considerable extent the practical use of welding. The company was later taken over by BOC becoming BOC–Murex, which is now also part of the ESAB group.

By the 1930s the concept of shielding the metal-arc was well established and in fact the USA was by then referring to 'shielded metal-arc' to distinguish the covered electrode process from its predecessor the bare-wire process. In Europe where the bare-wire process became obsolescent more quickly the process became simply metal-arc and then 'manual metal-arc' to distinguish it from the subsequent development of automatic welding.

Up until the mid-1930s metal-arc welding had only been carried out by the manual process in which a bare or coated wire of about one-third of a metre (12–14 in.) length was gripped in an insulated holder manipulated by the welder. The next step was to automate the concept. By using an automatic feeding device two important advantages were achieved. First the process could be made continuous and secondly thicker wire, giving higher welding currents and faster welding, could be used. Various ingenious methods were developed for supplying the protective flux shield to the molten pool. The most important of these was a blanket of a fused powdered flux administered through a tube ahead of the welding wire. The arc was able to work satisfactorily out of sight beneath this powdered flux and the process was therefore called 'submerged-arc'.

Also in the 1930s experiments began with the shielding of the arc by inert gases. This development took place in the USA where naturally occurring helium was available in reasonable quantities and had already been used for filling airships and blimps. The first gas-shielded process used a tungsten electrode shielded by inert gases, either helium or argon. This is now known as gas tungsten arc (GTA) or TIG. Within a few years a second inert gas-shielded process had been developed in which the tungsten electrode was replaced by a continuously fed metal wire. This was the gas metal-arc (GMA) or MIG process. Used initially with helium or argon for welding aluminium or other non-ferrous metals its use on steels had to await the development of special electrode wires with low carbon contents and a carefully balanced addition of deoxidising agents. This was necessary to prevent the occurrence of porosity within the weld.

The next development was to use a more readily available gas to shield the gas metal-arc than argon or helium. The obvious choice was carbon dioxide but there were two initial difficulties to be overcome. First the transfer of metal from the wire to the pool through the arc lacked directon, was indiscriminate and even explosive, and secondly the weld developed severe porosity as a result of the reaction of carbon in the molten pool with oxygen from dissociated carbon dioxide in the arc atmosphere.

A great improvement in the behaviour of the arc and metal transfer was achieved by the discovery of 'dip transfer'. This required the use of a short arc with the consumable electrode wire being melted into the molten pool through a rapid succession of short circuits during which the arc is extinguished. To achieve this a considerable development was necessary in power sources. Previously the role of the power source was predominantly to convert, with no great refinement, the high-voltage low-current input from the mains to a low-voltage high-current output suitable for welding. Power sources for dip-transfer had to be modified to provide sufficient stored energy to fuse the short circuit and re-start the arc several times a second. The dip-transfer or short-circuit technique, which could only be carried out at low welding currents, enabled sheet material to be welded and, because of the small molten pool, also enabled welding in position. Work on improving power sources continued for many years to reduce spatter and improve performance and was responsible for major changes in power plants, some of which are only now coming into widespread use. Power sources today have had the benefit of electronic and microprocessor control which has allowed many alternative methods of modifying the melting process at the tip of the electrode.

The solution to the porosity problem was provided by a wire containing relatively large additions of silicon, manganese and other deoxidants such as aluminium, titanium or zirconium. In retrospect the wire composition changes required seem obvious but at the time many of the results seemed puzzling and there were many false starts.

With the use of carbon dioxide as a shielding gas the term CO_2-welding came into use and more recently a modification of MIG to MAG, meaning metal active gas. While the development of CO_2-welding was taking place the parallel development of MIG welding for non-ferrous metals had pushed up the usage of argon. Gas producers were encouraged to separate this gas in oxygen plants and as a result the availability of argon improved considerably. Argon is now used with carbon dioxide or with oxygen in a range of mixtures for welding a variety of mild, low alloy and high alloy steels. The addition of argon or helium to the gas shield confers smooth running to the arc and reduces spatter. These are characteristics greatly in demand for arc-welding robots, a development which in the last few years has ensured a continuing increase in the use of MIG welding.

The development of processes using continuous wire feeders made it possible to devise wires which were hollow and contained inside the fluxes and other agents which had long been used on the outside of the electrodes for manual metal-arc welding. In fact, the cored wire was originally conceived as a solution to the carbon dioxide porosity problem just mentioned. At the present time cored wires are beginning to find increasing use on their own merits. One of these is the ability to alter the composition of the deposited metal at will, without having to go to the trouble of having special casts of steel such as is necessary with solid wires. Cored wires are used at higher welding currents than the solid wires used in the dip transfer technique. Small diameter gas-shielded cored wires are still being developed with higher deposition rates than their solid wire counterparts. Certain types contain gas making chemicals and other deoxidants which allow them to be used without a gas shield. These are the so-called self-shielded wires which allow considerable simplification of the equipment because gas cylinders, pipes and nozzles are not needed and therefore find their main use in site or open-air welding.

In the early days of welding there was concern that the weld metal would be the weakest part of the joint and that premature failure would occur here. Actually any reasonably sound weld metal has both a yield and ultimate strength in excess of those of the parent metal in which it is surrounded. This higher strength of a weld metal with a nominal composition similar to the parent metal is a result of metallurgical effects brought about by rapid cooling and straining. It is only in recent years when the strength of steels in common use has been raised that the electrode manufacturers have had to increase the strength of the weld metal to match that of the parent metal. Even weld metal in non-ferrous metals, for example aluminium alloys, is considerably stronger than metal of the same composition cast in the normal way.

Fusion welding

The arc-welding processes are undoubtedly the most important 'fusion welding' processes. There are at least 35 different welding processes (see the Glossary of Terms) and in the majority of these heat is used to melt metal to form a bridge between the parts to be joined so that on

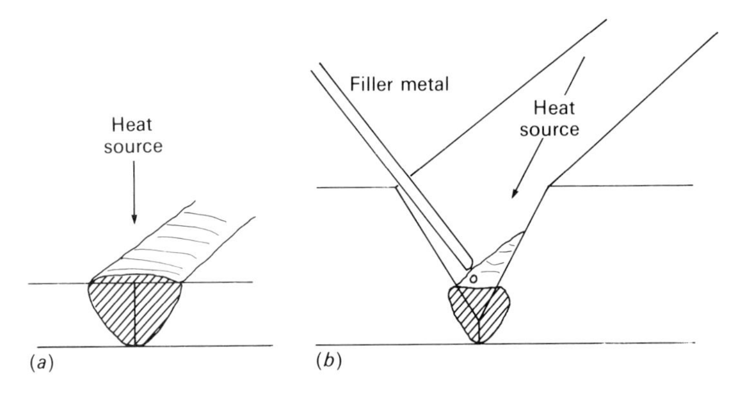

Fig. 1.2 *Welding* (a) *with and* (b) *without filler metal.*

removal of the heat source and solidification, the parts are united. Figure 1.2 shows that the metal which is melted can come exclusively from the parts to be joined when they are relatively thin, or may include extra metal known as filler metal, when thick metal with a prepared edge is to be joined. The heat to melt metal in the joint can be provided by a variety of heat soures such as: a carbon arc, an oxy-fuel gas flame, flux-shielded arcs, gas shielded arcs, electron beams or laser beams. In every such case part of the joint in the workpiece is metal which has been melted or 'fused', hence the name 'fusion welding' for these welding processes.

Non-fusion welding

Although this book is concerned only with fusion welding processes it is helpful to consider briefly those welding processes which do not qualify for this description. They are sometimes referred to as 'solid-state' methods or processes employing pressure, in which the two sides of the joint are brought into intimate atomic contact either by mechanical deformation or atomic diffusion or by a combination of both.

Ductile metals such as copper or aluminium can be joined by mechanical deformation at room temperature. For most metals, however, heat is used to render the workpiece ductile. In the forge welding of antiquity deformation was provided by hammering. Friction welding employs a rubbing action at the interface to generate heat and to provide the mechanical deformation. In diffusion bonding clean surfaces heated under enough pressure to cause local deformation are united by atomic diffusion across the interface. There are other processes, e.g., flash welding and MIAB in which there is melting, but because pressure is employed the molten metal is squeezed out leaving the joint free from cast or 'fused' metal. Finally, the electric resistance welding processes such as spot or seam welding, although they involve joining through a nugget of fused metal, are never classed as 'fusion welds'.

There are a great variety of non-fusion welding processes and they tend to be more specialised in their application than the fusion welding methods. For example, spot welding is used only for joining overlapping sheets and friction welding is suitable for bar and tube stock. Non-fusion processes are also methods which are mechanised and in which the manual dexterity of the welder plays no part. One advantage which these methods have over the fusion methods is that because there is no, or very little, cast metal in the joint, they are generally more suited for joining dissimilar metals. The fact that the joint comprises only metal from the workpiece, unlike the fusion processes where filler metal is often added, is both an advantage and a disadvantage. It is in fact helpful sometimes to be able to add extra metal to a joint so that changes may be made in the composition of the weld metal to give the required properties.

The fusion welding processes based on the welding arc have evolved and developed great refinement. As a group they are responsible for the overwhelming majority of all welds made and their variety makes it possible for the user to select the process and technique to suit the application and economics required. In spite of the developments which have been made to date the search is still on for new heat sources of which electron beam and laser are the most recent examples.

Power-beam processes

In the early 1960s the advent of nuclear power and the availability of reactive metals such as titanium and zirconium called for new standards of quality and protection from the atmosphere while welding. Considerable effort was put into developing tungsten arc welding for reactive metals, often with the use of torches inside chambers and glove boxes. It was against this background that the electron beam process made its appearance. A stream of electrons, from what at the time was considered a high-power electron gun, was focused on the joint to produce fusion. Because the electron gun required a vacuum in which to work the whole operation was carried out in a vacuum chamber, supplying as well the necessary protection for the workpiece by total exclusion of the atmosphere. It was quickly found that the electron beam process had other advantages that were of importance for general engineering and not merely for the esoteric applications originally envisaged.

The energy concentration which can be reached in the focused spot of an electron beam is so intense that it is capable of boring its way through a workpiece and welding can be carried out rather like drawing a hot wire through butter! The deep narrow molten pool which results and the high speed of welding can almost eliminate that bane of the arc welder – distortion. Distortion is caused by the localised heating and cooling associated with the heat source and molten pool. The smaller the molten pool the less the distortion. Edge preparation is unnecessary with electron beam, and square close butt joints are used even in thick plate. Another attraction is the range of operation of the process which can join delicate parts with minimum heat effect but can also be operated at higher power to join in a single operation metal of several hundred millimetres thickness. At the top end of its range electron-beam welding is a potential competitor to submerged arc welding.

The laser is another heat source capable of generating in a focused spot the same levels of energy intensity as the electron beam. When it was developed to the point where it could be shown to penetrate metal like an electron beam it was naturally examined with the same enthusiasm as had been the electron beam, the more so since laser beams can operate in air without the need for a vacuum chamber. Both laser and electron beam processes use relatively costly equipment and for this reason their penetration into industry has been slow and selective, the main applications being in aerospace where quality and freedom from distortion are the requirements and the automobile component industry where welding speed is often the most pressing need.

Cutting

While heat can be made to fuse and unite, it can also be employed to separate, that is, perform a cutting operation. The earliest arc welding patent for the Benardos carbon-arc process also mentions cutting. This was achieved by forming a molten pool and allowing it to fall out by gravity. The results of thermal cutting in this way are rough and generally unsatisfactory but great improvements result from combining the heat source with a jet of gas, usually oxygen or air. An air jet with a carbon arc is widely used for gouging the surface of steel. The arc melts a pool on the surface and the air jet blows it away. An oxygen jet with an oxy-acetylene heat source was the most commonly used thermal cutting process for steel, although today oxy-propane has become widely used. Here the oxygen jet oxidises the steel providing additional heat through an exothermic reaction and also blows the oxides through the cut (see Fig. 1.3). In plasma cutting an inert gas tungsten arc is constricted by making the arc pass through a water-cooled nozzle and the inert gas thus concentrated forms the jet which ejects the molten metal from the cut. A recent development has allowed air to be used in

Fig. 1.3 *Oxy-fuel gas cutting steel plate.*

the plasma torch for cutting thin sheet steel. This has only been possible by finding an electrode material more resistant to oxidation than tungsten. Gas-jet laser cutting uses a laser heat source and a jet of oxygen for ferrous metals or an inert gas or air for non-metals. This range of thermal cutting processes is now offering severe competition to the traditional methods of cutting shapes using the blanking die, shear, punch and nibbler. Most cutting processes, including oxy-fuel gas cutting, are available with computer controlled profiling equipment which can access standard shapes in memory and select the sequence of cutting and the arrangement of shapes to provide the most economical use of material.

Summary

The most common and widely used welding processes are those which employ fusion, that is, a heat source is used to cause local melting which allows the parts to be welded to flow together. It may be necessary to add extra metal to the joint and this is known as filler metal. When the heat source is removed the molten metal solidifies, thereby consolidating the weld. The most widely used heat source is the electric arc in which the arc is struck between the workpiece and an electrode. If the electrode is made of a wire of a similar composition to the workpiece (parent metal) it will melt at a rate depending on the welding current and wire diameter and composition fusing the parent plate and providing filler metal for the joint.

Protection is required for the hot metal from the atmosphere and this may be achieved by covering the wire in a flux as with the manual metal arc (or stick electrode), or by shielding the arc with a gas as in gas metal-arc welding. The flux is a convenient way of adding elements such as silicon and manganese, which are necessary to control porosity and provide the required strength in the weld metal. When a gas is used for shielding these elements are added to the filler wire.

The shielding gas may be inert as in MIG welding when argon or helium or mixtures containing these gases are used, or it may be an active gas such as carbon dioxide. When carbon dioxide is used even when mixed with some inert gas the process is limited to use on steels and may be called MAG welding. MIG and MAG welding use a continuous electrode wire because electrical contact can be maintained through the surface of the wire. Continuous welding is also possible when the wire is made in a tubular form and contains flux. This is called cored-wire welding and it may be carried out with gas shielding, or the flux in the core may be so constituted as to make extra gas shielding unnecessary. A mechanised form of arc welding known as submerged arc welding uses a consumable electrode in which the arc is submerged under a layer of powdered flux.

If the electrode is made from a metal such as tungsten which does not melt during welding the process is called gas tungsten arc or GTA or TIG welding, and protection from the atmosphere is provided by an inert gas. One of the earliest welding processes was gas welding employing a flame in which a fuel gas, e.g., acteylene, is burnt in oxygen. As with GTA welding filler metal must be added separately, in this case by hand. Other heat sources using electron or laser beams, the so-called power beam processes, are being used increasingly and have the advantage of a high intensity heat source giving deep penetration welds allowing welding with a minimum of distortion.

Before discussing the various fusion welding processes there are certain general principles concerning the processes and welding metallurgy which will be dealt with in Chapters 2 and 3. A list of welding terms appears in the Glossary on page 209.

Chapter 2

Principles of arc welding

There are a number of basic principles which are common to many of the arc, gas and power-beam processes considered in this book. To avoid repetition and to provide a broader information base for further understanding these are considered now in sub-sections of this chapter covering the following:

1. The nature of welding arcs, how they are initiated and maintained.
2. The different power source types that are commonly used to control the welding arcs, and some of the terminology used to classify them.
3. The role and function of welding consumables.
4. The properties of weld metal and some of the common weld metal defects.
5. The effect of the welding process on the workpiece or parent metal in terms of properties, defects and distortion.
6. The metallurgical specifications and procedures for materials that are frequently welded. This topic is covered in Chapter 3.

Each of these topics is a subject in itself and there is much more detail to them than can be presented here. However, from the point of view of the user, power source design and arc physics are often of secondary interest in much the same way as a car driver can choose a car and learn to operate it safely without having to understand fully the design of the engine and the functioning of the carburettor. Lack of knowledge or understanding of consumables and material selection can under certain circumstances, however, lead to serious problems or even catastrophic failure. It is for this reason that metallurgical aspects are explained in some detail whereas only the basic principles of equipment design are described.

It is hoped that a general understanding of the matters mentioned above and presented further in Chapters 2 and 3 will assist the reader in considering the main fusion welding processes to be dealt with in the subsequent chapters. A section on sources of information and a bibliography which appears on pages 224–7 has been included to enable readers to follow up points of detail should they wish to do so.

The welding arc

The nature of the arc

An electric arc is an electrical discharge between two electrodes. The energy supplied has to be sufficient to allow the discharge to take place through an ionised gas known as 'plasma'. Extremely high temperatures, of tens of thousands of degrees, can exist in the core of the arc plasma. When used for welding, one of the electrodes is generally the workpiece and therefore a plane surface while the other pole of the arc is the electrode which approximates to a point. Under these conditions the arc spreads from the electrode forming a bell shape. The flow of current in the arc results in a magnetic field which envelopes the arc and tends to compress it, the so-called 'pinch' effect. Because of the way the arc expands from the electrode to the workpiece there is an axial component of the pinch effect which causes the hot ionised gas in the arc to be set in motion from the electrode to the work. This is sometimes called a plasma jet and it is responsible for the depression which forms in the surface of molten pools at high welding currents. The most important result of the electro-magnetic forces in an arc and the plasma jet is the way these forces can detach molten metal from the end of the electrode and transport it to the molten pool – even against the influence of gravity.

An important characteristic of a welding arc is whether or not the electrode is melted. With high melting-point electrodes such as carbon or tungsten, so-called refractory electrodes, the electrode is not melted. These electrodes are said to be 'non-consumable'. When a lower melting-point metal is used, e.g. when the electrode is the same as the parent metal, the end of the electrode melts and molten droplets can be detached and transported to the molten pool in a manner which depends upon the metal and the nature of the arc atmosphere. This sort of electrode is described as 'consumable'. If a consumable electrode is shielded by flux as in manual metal-arc welding it is necessary to arrange the composition of the flux covering to have the desired influence on the transfer of metal in the arc. As we shall see in later sections, however, the flux has other functions to perform and the final choice of flux composition is always a compromise. With gas-shielded consumable electrodes the way in which metal is transferred for any given metal depends on the gas composition, the level of the welding current and also the shape of the current waveform.

The current supplied to an arc is generally straightforward AC or DC but some modern welding power sources have square wave output or allow the welding current to be pulsed beween high and low values. This is done at frequencies of 10–100 Hz to control the detachment of droplets from the welding wire in gas metal-arc welding or at frequencies in the range one every few seconds to 5 Hz to control the molten pool in tungsten arc welding. With the latter welding proceeds by making a series of overlapping weld nuggets which provides good control of penetration and flow of heat into the surrounding metal enabling easier welding.

Initiating an arc

Arcs cannot be merely switched on and off. The plasma discharge has to be established usually by touching the electrode on the workpiece and then withdrawing it so that the discharge is lengthened to that required for welding. 'Drawing' an arc in this manner is sometimes inconvenient with automated gas tungsten arc equipment and various techniques such as high-frequency discharges or spark generation are used. High-frequency discharge starting is used occasionally with gas metal-arc processes but generally the small-diameter electrode wire used with these processes will fuse near where the tip contacts the workpiece with the initial surge of current. It is not necessary, therefore, to withdraw the electrode to start the arc as the arc gap is created by a mechanism like the blowing of a fuse.

The initial arc formed in the manner just described will not grow into a stable welding arc if the power source cannot deliver the current required. Power sources must therefore have appropriate dynamic and static responses. One of the reasons why batteries were the chosen power source in the very early days of arc welding was probably that arc starting was easier. Once the arc has been established it has to be maintained during the current zero period if it is an AC arc and during inadvertent arc-length fluctuations when welding is in progress. The most important factor in this is the voltage which the power source provides at the moment of arc extinction. With manual metal arc electrodes the voltage at which the arc goes out and the voltage at which it restarts depends on the composition of the electrode coating. Certain compounds, for example, sodium or potassium silicates, enable the arc to run at lower voltages which in turn mean that the power source can have a lower open circuit voltage. Electrodes with a covering containing cellulose produce an atmosphere round the arc which contains hydrogen which raises the voltage required across the arc. This provides a hotter deeply penetrating arc but with the concomitant requirement for a higher open circuit voltage from the power source.

Arc maintenance

Once the arc has been ignited it has to be maintained by the power source providing an appropriate voltage at the arc. All other things being equal the voltage required by an arc depends directly on its length or the gap between the electrode and the molten pool in the workpiece. Welding conditions should be stable and a steady arc length is generally required, except for certain welding techniques, to be described later, in which the arc length is deliberately caused to fluctuate. When fluxes are employed these can influence the stability of the arc and compounds are often added to make arc recognition easier after the voltage reversals in AC welding.

Effect on the workpiece

As will be seen later the rate at which energy is applied to the workpiece by the welding arc has an important influence on the metallurgical properties of both weld metal and heat affected zone. Heat input is generally defined as kilo-joules/mm (kJ/mm or kJ/in) of weld or weld run, i.e., current × arc volts divided by welding speed. Actually a proportion 10–40 per cent of the energy generated in the arc never reaches the workpiece because of radiation losses in the arc column, losses in spatter and also when the electrode is non-consumable through losses to the welding torch. The magnetic field around the arc increases with welding current and tends to compress the arc increasing the temperature of the arc core and the forces such as plasma jets within the arc. These magnetic fields are susceptible to deflection by other magnetic fields as a result of the welding current flowing in the workpiece or residual magnetism in the steel being welded. When welding automatically, it is sometimes desired to use magnetic effects to deliberately deflect the arc in a cyclical manner, e.g. for weaving, and this is done by placing an electro-magnet close to the arc. All these effects are manifestations of what is called 'arc blow' which is of concern mainly with DC welding.

When the arc plays on a workpiece the roots of the arc tend to seek out and consume those areas having a high emissivity such as oxides. This arc cleaning process occurs particularly when the electrode is positive and is a phenomenon made use of when welding metals with tenacious oxide films such as aluminium alloys.

Power source principles

Input power from the mains is usually high voltage low amperage, but welding demands high amperage at a low voltage. Where mains supply is not available or convenient, power can be generated from engine-driven dynamos. To obtain the required control of the arc as a heat source and to ensure suitable metal transfer, power sources are arranged. to have a variety of different characteristics.

Control of output

There are two main functions to any welding power source. First, it must produce a flow of current at a suitable voltage, and secondly it must control or regulate the output at the level required for welding. In the early days of arc welding all welding power sources were of the constant-current type, that is the variation in current drawn from the power source as a result of the normal arc voltage variations in manual welding was minimal. A plot of the current drawn vs. voltage for such a power source is of the type shown in Fig. 2.1 and is known as a drooping characteristic. The intersection with the abscissa indicates the current drawn on short circuiting the power source. These constant-current power sources were largely used for manual metal-arc welding (see Chapter 4). Also shown on the diagram is the volt-amp curve for an arc. The intersection of the two curves defines the operating point.

With gas-shielded metal-arc welding, see Chapter 5, in which small diameter electrode wires are fed into the arc at a pre-set constant speed a constant potential or constant voltage power source is generally preferred, the welding current being then controlled primarily by the wire feed speed. This is because with this welding process control of the arc length is achieved by influencing the burn-off rate by short-term

Fig. 2.1 *Volt/amp characteristics for an arc and a constant current power source.*

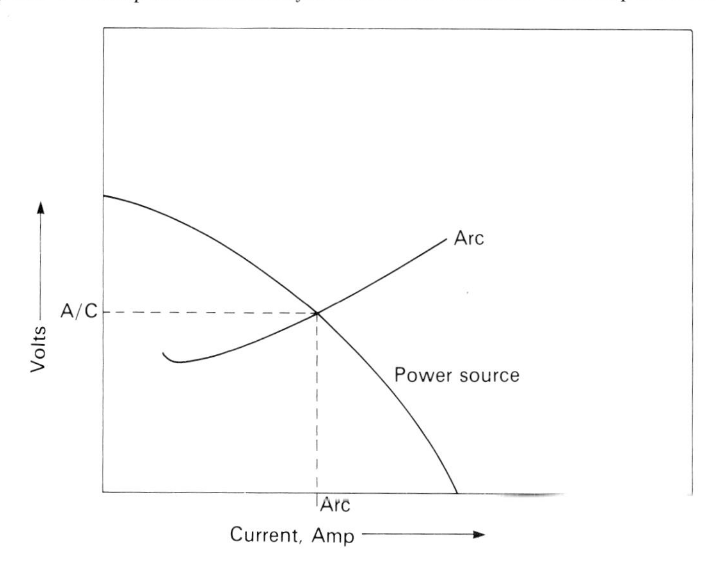

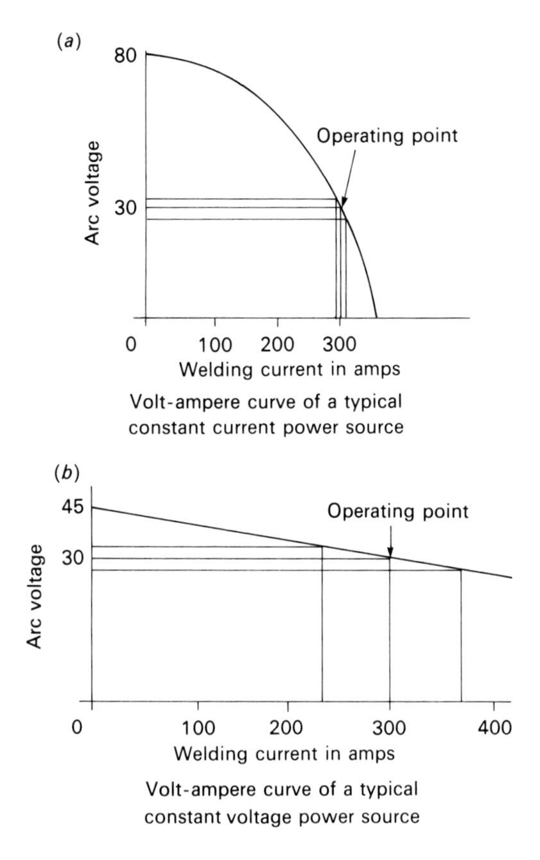

Fig. 2.2 *Comparison of constant current and constant voltage power source characteristics showing the greater current swing with the constant current power source for the same voltage change.*

variations in welding current. The larger the voltage/arc length changes resulting from poor control in positioning the electrode, the greater are the current swings in response. Gas-shielded metal arc welding is said to possess inherent self-adjustment of arc length, something which will be discussed in more detail in Chapter 5. Constant current and constant voltage output characteristics are illustrated in Fig. 2.2 (a) and (b).

Power source types

Apart from batteries the earliest welding power source was the DC generator – a dynamo driven by mechanical means or from an electric motor. With a generator the welding current control can be by means of an external resistor or choke but it is usually achieved by additional coils within the generator itself. The majority of DC power sources in use today, however, are transformer/rectifiers. These will be discussed later after the section on AC welding with transformers.

AC sources
AC power sources are almost without exception of the constant current or drooping type and fall into five main categories as follows:

1. Moving iron.
2. Moving coil.
3. Tapped reactance.
4. Magnetic amplifier.
5. Electronic control.

The AC power source comprises a transformer, the function of which is to reduce the voltage from that of the mains to what is required for welding and also to isolate the mains from the welding circuit, and a regulator to control the current drawn from the transformer. The difference between the various AC power sources is in the way this regulation is achieved.

Moving iron transformers In this type of transformer the reactance of the power source is changed by moving an additional iron core in or out of the magnetic field of the transformer. Remote control can be obtained by having a motor to drive the iron core in and out of the windings.

Moving coil transformers In these transformers the central core is static and the spacing of the coils is varied to achieve current control. Remote control is by motor moving the coils.

Tapped reactance With this power source the output from the transformer is fed through a tapped reactor or choke in the secondary circuit.

Magnetic amplifier These power sources are also known as saturable reactor or transductor power sources. The performance of the reactor or choke is altered by having a DC winding as well as the AC output winding. The DC winding is fed from a subsidiary winding in the main transformer, the AC from which is rectified and passed through a variable resistance. Increasing the flow of DC reduces the reactance of the choke. This type has the advantage that it is particularly easy for control to be carried out remotely from the transformer by having the variable resistor on a separate lead.

Electronic control Modern electronic power sources using either thyristor or transistor technology can produce both AC and DC outputs.

They offer the capability to produce square rather than sinusoidal AC waveforms and permit precise setting of welding current. Feedback control technology ensures absolute reproduceability against mains or temperature variations for example.

Most AC power sources incorporate power factor correction capacitors to maximise the efficiency of the transformer and improve the inherently lagging power factor.

DC sources

There are three main types of DC power source each of which may have either a drooping characteristic or a flat characteristic (see later section) as follows:

1. Transformer rectifier.
2. Electronic.
3. Inverter.

Transformer rectifier Transformer-rectifier DC power sources generally comprise three parts – the transformer, the current control device and the rectifier which converts the output of the first two parts from AC to DC. The current control devices are similar to those used for AC power sources and the rectifiers are conventional electrical devices (usually silicon type). Domestic supply is single phase, but more usually for industrial use three-phase power input is employed, the latter giving an output which has lower voltage ripple. Some ripple in voltage is inevitable but may be reduced by the insertion of filtering components in the output circuit. By using internal switching it is possible to make a transformer–rectifier power source provide both AC and DC output. As heat is generated in the rectifier stacks and other components of a transformer–rectifier DC source it is usual to enclose a fan for cooling within the casing.

Electronic Developments in electronic devices over the last twenty years have provided the opportunity for new dimensions in the control of power source output. Both thyristors and transistors are now used in power sources and permit steady and pulsed DC and square-wave AC. Such power sources can be used for gas tungsten arc, gas metal-arc, and manual metal-arc. As stated previously one of the main advantages of electronically controlled power sources over their conventional counterparts is that variations in mains input voltage can be compensated to provide a consistent output. This is most important if precise output conditions have to be guaranteed to ensure reproducible weld quality.

Inverter The use of an inverter enables the size, weight and potentially the cost of power sources to be reduced. Use is made of the fact that as

Fig. 2.3 *An inverter power source with an output of 200 A with a 100 per cent duty cycle and 315 A with 35 per cent duty cycle.*

the frequency at which a transformer operates is raised so its size may be reduced. The incoming AC is rectified and then chopped by the inverter to produce a high-frequency AC which is then transformed to a voltage suitable for welding. Finally, this low-voltage, high-frequency output is rectified to provide DC for welding. Figure 2.3 illustrates that the reduction in size and weight can be quite dramatic. The reduction in size and weight is apparent from Fig. 2.3 in which the power source is being held by hand.

Duty cycle

Heat is always generated within a power source because of its internal electrical resistance and magnetic characteristics. Such heat which is undesirable and in the extreme may result in serious damage to the equipment can be minimised for example by being more generous in the cross-sectional area of the conductors or transformer core. This puts up the cost of the equipment and so a compromise is necessary, for which purpose the equipment designer uses the concept of 'duty cycle'. The heating effect depends on the level of current and the duration and frequency of use. Power sources used for manual welding which receive only intermittent use and for relatively short periods can be run at higher working currents than if they were used with automatic processes when the periods in use can be more lengthy and frequent. The definition of duty cycle and its use is the subject of national and international standards but is usually expressed in terms of the 'rated' or permissible current for a given duty cycle or period of continuous operation.

Volt–amp characteristics

The most common way of characterising a power source is by the relationship between the current delivered and the voltage across the output terminals, known as the volt-amp characteristic. Although a most useful description, however, this does not characterise a power source fully as the relationship is a static one since the measurements are made under steady-state conditions. Figure 2.4 shows the characteristic for a typical 'drooping' or constant-current power source together with the characteristics of a welding arc at several different arc lengths. The point where the arc characteristic at the appropriate voltage crosses the characteristic for the power source indicates the operating conditions and the way arc current will change with changing voltage. It will be observed that if the power source characteristic is much flatter than shown, i.e., if it is a constant voltage power source there will be a proportionally greater swing in current for the same voltage change. This is why a constant-current power source is usually selected for manual metal-arc welding where consistent voltage may be difficult to achieve.

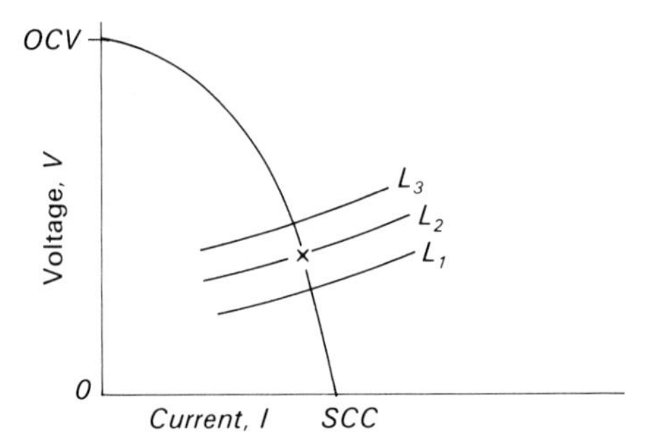

Fig. 2.4 *Arc and constant current power source characteristics for arcs of three different lengths. L3 longer than L2 longer than L1. OCV = open circuit voltage. SCC = short circuit current.*

Inductance

Mention has been made of the use of a reactor or choke in AC circuits to limit current. The effect of inductance in a circuit is to delay the rise and fall of current in a coil in an exponential manner as a result of the building up and decay of the magnetic field. The result is that the current lags behind the voltage. With a pure series resistance voltage and current are in phase. No circuit, particularly in welding equipment, however, is either purely resistive or purely inductive.

In AC power sources and the AC stages of a DC power source the variable reactor limits the current which is drawn at any voltage from open circuit to short circuit. Inductive reactors are, however, used in the DC stages of power sources where they would behave as resistors if the current were steady. Welding current is never steady, however, and the reactor in this case is there to smooth and dynamically control the output. It prevents the explosive fusing of the bridge between pool and electrode on a short circuit and supplies, because of its stored energy, extra volts to help re-ignite the arc after a short circuit has been cleared. These functions are particularly important in the dip-transfer or short-circuit welding of ferrous materials.

Consumable principles

The term consumables is applied to the supplies which are consumed during the making of a weld. It includes the oxygen and fuel gases for gas welding and the welding wire and flux which may be used. With manual metal-arc welding it is the electrodes used and with submerged-arc welding the wire and the flux. With gas-shielded welding it includes the shielding gas, argon, helium, carbon dioxide or gas mixtures, as well as filler wire and electrode wire. It also includes spares, such as nozzles, and the electricity used in welding. Usually, however, the term is restricted to those items mentioned above which are responsible for controlling or influencing the metallurgical quality or mechanical properties of the welded joint.

A primary function is to protect the hot and molten metal in the arc and pool from the atmosphere, the danger being from both nitrogen and oxygen which react with the molten metal. Oxides impair the flow and fusion of metal and influence the formation of inclusions; they also use up certain alloying elements. Nitrogen forms nitrides, which can be a cause of brittleness, as well as porosity. In gas-shielded welding with an inert gas like argon, oxygen and nitrogen are excluded passively. Some oxides are always present, however, so it is necessary to flux them away with a deoxidant in the electrode such as silicon, aluminium or titanium with steel or silicon, zinc or phosphorus with copper alloys. Aluminium oxide cannot be reduced chemically so the gas shielding when welding aluminium alloys must be as effective as possible. Carbon dioxide shielding, often used with steels, excludes nitrogen at the expense of allowing oxides to form and consumables containing rather higher amounts of deoxidants are normally required. Protection from the atmosphere with a process employing a flux, such as manual metal arc, is twofold. From a surface protective layer of molten flux and also from gases formed from components in the flux which are decomposed by heat.

The way the deoxidant is applied depends on the welding process. With gas-shielded welding it will be a metallic element in the filler or electrode wire but with manual metal-arc there will be metallic deoxidants in the electrode covering. Occasionally as with some copper-based alloys the deoxidant is also present in the parent metal. This is particularly useful when welds are being made by gas tungsten-arc or electron-beam processes during which it is perhaps difficult or economically undesirable to add a filler metal.

Frequently alloying elements present in the parent metal fulfil a deoxidising function, e.g. copper alloys with small alloying contents of zinc or silicon. Even with ferrous metals, where it is not usual to rely on alloying additions to fulfil a deoxidising function, many complex reactions take place between alloying elements and the consumables or atmosphere above the molten metal. Elements such as carbon, silicon, manganese, titanium, aluminium and chromium are readily oxidised and a proportion of the initial content of these elements will be lost during welding if the filler wire and flux do not provide a balance to the equation.

Consumables have a powerful effect upon the way metal is transferred from the electrode to the molten pool. With a manual metal arc electrode the presence of a large amount of mineral rutile in the covering causes smooth transfer in small droplets but with a high calcium carbonate content transfer takes place in large globules which short circuit to the molten pool into which they become incorporated by an action known as 'bridging'. When using gases such as argon or helium for shielding in steel welding it is necessary to add small amounts of oxygen or carbon dioxide to give smooth spatter-free transfer in small droplets. Such additions are harmful when welding aluminium alloys but with this material the smoothness of metal transfer and the shape of the penetration bead can be modified by using mixtures of argon and helium.

With a welding process using a flux there will be a flux residue called slag left on the surface of the pool. The viscosity, surface tension, melting-point and other characteristics of this slag influence the final shape of the surface of the weld bead, the way it blends into the parent metal and the ease with which welding may be carried out in positions other than the flat or gravity position. It is also important that when welds are being made in deep grooves that any flux residue can be easily removed to allow further weld beads to be laid on top.

These considerations of metal transfer, ease of welding, control of weld bead composition, weld properties and of course cost are important in selecting consumables. A compromise is often necessary as will be explored further in subsequent chapters dealing with particular welding processes.

Weld metal properties and weld defects

The molten pool and dilution

It would be ideal if the properties of both the weld metal itself and the zone alongside affected by heat had exactly the same properties and characteristics as the parent metal. This is not possible, however, because weld metal is cast whereas most parent metals are used in the wrought form. Wrought materials almost always have higher strength, ductility and toughness than the comparable material in cast form. Weld metal is, however, a miniature casting which is rapidly cooled and its properties often approach those found in wrought material. This is particularly the situation with ferrous materials but the property match is less satisfactory with some non-ferrous metals such as aluminium and copper alloys.

Because of the electro-magnetic forces within an arc, mentioned earlier, molten pools are set in motion internally with a variety of flow patterns depending on the joint type, welding current and the angle the torch or electrode makes with the weld line. This turbulence results in a uniformity of temperature and composition within the liquid metal apart from the hot spot in the immediate vicinity of the arc root. The final composition of the weld metal is a result of a mixture of the electrode or filler metal melted and the parent metal which is melted. The metal deposited from the electrode or filler wire is said to have been 'diluted' by melted parent metal. When no metal is added at all and the weld metal consists entirely of parent metal the dilution is said to be 100 per cent. With manual metal-arc welding the root run may have 30 per cent dilution with the subsequent runs having slightly less. As a result of the uniformity of the weld metal it is possible to calculate its composition if the analysis and relative proportions of melted parent metal and electrode can be estimated. This is often done from an examination of a cross-section of the weld as Fig. 2.5 shows. Such calculations which merely involve simple proportion, are important when a filler metal or electrode of different composition from the parent metal are used, as in dissimilar metal joining, surfacing mild steel with stainless steel, or in the welding of aluminium alloys. It may also be necessary to consider dilution if the

Fig. 2.5 *Estimating dilution from weld geometry.*

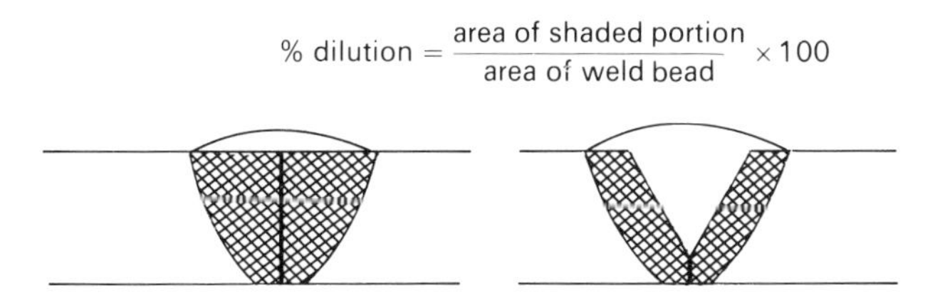

parent plate has a high sulphur content or if it contains aluminium which if taken into the molten pool may affect the oxygen content and impair the toughness of the deposit.

Weld metal structure

In carbon and carbon-manganese steel weld beads the columnar grains are outlined in ferrite and there are frequently laths of ferrite growing in from the grain boundary. This type of structure has poor toughness and if it is necessary that it be removed the usual method is a normalising heat treatment. Where a multipass weld is made, however, each weld bead is heat treated by the bead which follows. Metal which is heated above the transformation temperature range crystallises to a finer equiaxed grain size. The depth to which this recrystallisation takes place depends upon many factors, including the interpass temperature, and it is unusual for complete recrystallisation to occur (see Fig. 2.6). The effect also refines the structure in any adjacent parts of the heat affected zone. A critical region in which toughness is desirable is at the toe of the

Fig. 2.6 *Recrystallisation of weld beads by subsequently deposited weld runs in a heavy fillet weld made by MMA.*

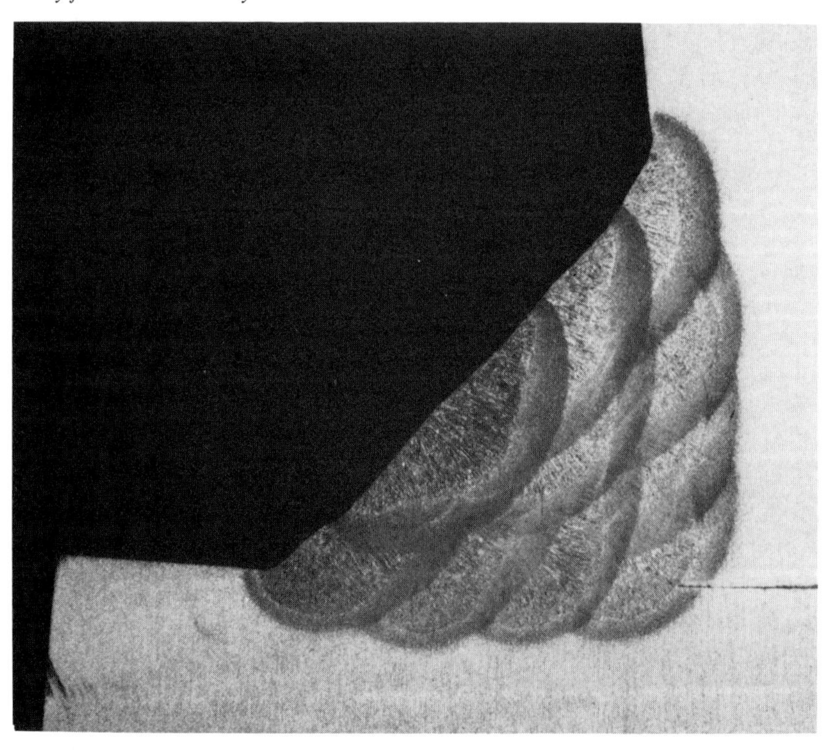

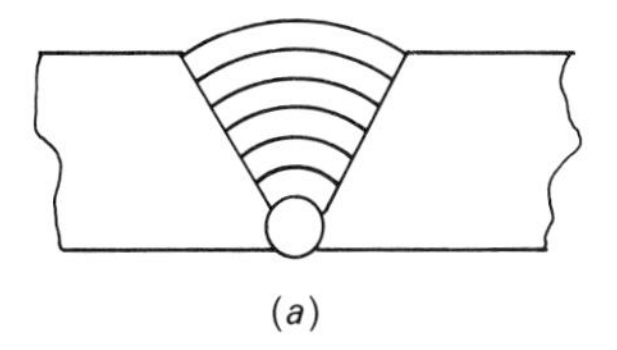

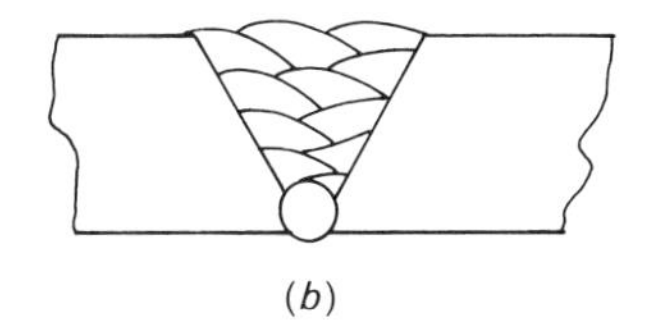

Fig. 2.7 *The techniques of* (a) *wide weave and* (b) *split weave.*

weld and this being the last metal to be deposited in a multipass weld may be without the benefit of recrystallisation treatment. Careful placing of the final bead or beads is necessary to ensure that grain refinement takes place where it is needed. Loss of toughness can occur in the HAZ of structural steels and this is associated with high heat input which causes grain growth and microstructural changes. Where toughness is important, as in structures which have to maintain their integrity at low operating temperatures, large weld beads are usually avoided by employing stringer bead or split weave techniques (see Fig. 2.7). With hardenable steels the rapid heating and cooling can create a hard layer of martensite alongside the weld bead. This also is reduced in hardness by the careful placing of the final weld beads. See page 35 for a further discussion of toughness.

Materials other than carbon steels which do not have a solid-state phase change do not show the grain refinement by successive weld beads. Other changes can occur in the reheated weld beads, however, such as the melting of any low melting-point grain boundary films to form incipient cracks. This can occur in the multipass welding of austenitic stainless steels.

Weld metal defects

Some of the defects which can occur in weld metal are:

1. Solidification cracking or 'hot' cracking.
2. Weld metal hydrogen cracking.
3. Porosity.
4. Slag or other inclusions.
5. Crater pipes and cracks.
6. Lack of fusion.
7. Poor weld bead shape.

Solidification cracking

Most steels can be welded with a weld metal of composition closely similar to that of the parent metal. Many highly alloyed steels and most non-ferrous alloys require electrodes or filler metal different from the parent metal because they have a wider range of temperature over which solidification takes place. This renders them susceptible to 'hot'

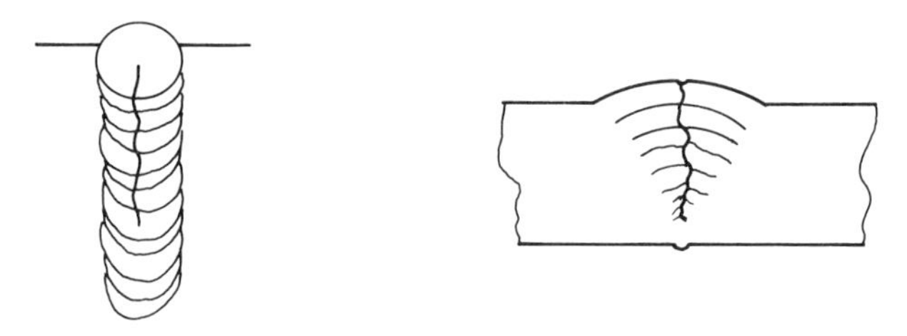

Fig. 2.8 *Centreline cracking in a deeply penetrating single deposit.*

or 'solidification' cracking which can only be avoided by welding with consumables specially chosen to reduce the solidification range and to provide so-called 'eutectic feeding' to fill incipient cracks. Hot cracking is also strongly influenced by the solidification direction taken by the solidifying grains in the weld (see Fig. 2.8). When grains from opposite sides grow together in a columnar manner low melting-point constituents and impurities can be swept ahead of the solidification front to form a line of weakness in the centre of the weld. Mild steel weld metal which inadvertently contains a high sulphur content can react in this manner and then centre line cracking can occur. Even when the sulphur content is normal the line of weakness may still exist to be opened up by strains during welding and it is for this reason that deeply penetrating weld beads are normally avoided.

Weld metal hydrogen cracking

This form of cracking occurs at temperatures around ambient and is more commonly observed in the heat affected zone. Hydrogen is introduced to the molten pool through moisture or hydrogen containing compounds in fluxes or on the surface of wires and parent metal, with the result that the molten pool and the solidified bead become a reservoir of dissolved hydrogen. In a steel molten pool the hydrogen diffuses from the weld bead to the adjacent regions of the heat affected zone which have been heated sufficiently to form austenite. As the weld cools the austenite transforms and further diffusion of hydrogen is almost eliminated. Hydrogen retained in this region adjacent to the weld bead can cause cracking, as will be discussed in the later section on the heat affected zone.

Porosity

This can occur in three ways. Firstly, as a result of chemical reactions within the molten pool, e.g., if a molten pool of steel is inadequately deoxidised oxides of iron may react with the carbon present to liberate carbon monoxide. Porosity can occur at the start of a weld run in manual metal-arc welding because at this point the shielding is not fully effective, see page 66. Secondly, by rejection of gas from solution as the weld solidifies, as occurs in the welding of aluminium alloys when hydrogen originating from moisture is absorbed in the pool and later released. Thirdly, by gas entrainment, e.g., the trapping of shielding gas in the root of turbulent molten pools in gas-shielded welding, or the gas evolved during the welding of the second side of a double 'T' fillet in primed steel plate. Most of these effects can be readily avoided and porosity is in any case not a defect which is particularly damaging to mechanical properties except when it breaks the surface. When this occurs it may provide a notch from which premature fatigue failure can occur.

Inclusions

With processes using a flux it is possible for particles of flux to be left behind to form inclusions in the weld bead. These are particularly likely in between successive runs of weld metal or between the weld metal and the face of the groove in the parent metal. The usual cause is inadequate cleaning between weld runs aggravated by poor welding technique with weld beads not blended smoothly into each other or the parent metal. As with porosity, isolated slag inclusions are not damaging to properties but aligned inclusions in certain critical positions, e.g., lying across the direction of stress might initiate more serious fracture. There are other forms of inclusion which are more common in non-ferrous and stainless steel welds than in structural steels. Oxide inclusions may be found in gas-shielded welds where the shielding has been inadequate and tungsten inclusions can result in gas tungsten arc welds from too high a current for the size of tungsten or by touching the tungsten on the work.

Crater defects

It has been mentioned that the grain structure in weld metal is generally columnar. These grains tend to grow from the grains present at the fusion boundary and grow away from the interface of the liquid metal and parent metal in the direction opposite to that of heat abstraction. A stationary fused spot is of course nearly hemispherical in outline but movement of the heat source produces a teardrop shape with the tail away from the direction of movement. The higher the speed the longer the tail. If the heat source is suddenly removed the molten pool freezes with a hollow or even a pipe in what is called the crater. The crater is also liable to contain star-shaped solidification cracks. Welding techniques at the end of a weld end are designed to correct this by momentarily moving the arc back along the weld or even gradually reducing the current without further movement.

Lack of fusion and poor weld shape
These are common defects which are easily avoided. The cause may be too low a welding current or an unsuitable welding speed. Both defects are discussed later.

The heat affected zone (HAZ)

General

No weld can be made without building up a thermal gradient in the parent metal. As has been indicated already the spread of heat into the parent metal is greatly influenced by the temperature of the molten pool and the speed of welding. High-power welding at high speed compresses the thermal gradient.

At a point in the HAZ just outside the edge of the liquid molten pool the temperature rises rapidly to a level close to that of the molten metal and declines rapidly to produce a quenching effect. In steels this region becomes austenite on heating and may contain the hard constituent martensite when it cools. This region develops coarse grains but further out, where the maximum temperature has not been so high, having entered the region above the transformation temperature but not reached the austenite region, the grain size is smaller. Further out still there is no change in grain size but the heat is sufficient to soften the steel and eliminate to some extent the effects of any cold working. Similar metallurgical effects are also observed in the HAZ after thermal cutting. With solution/hardening materials such as some of the aluminium alloys, the region close to the molten pool becomes effectively solution heat treated and will increase in hardness with time or a subsequent heat treatment at lower temperature which causes precipitation hardening. With materials which neither transform, like steels, nor solution harden, such as the heat treatable aluminium alloys, the effects of heat are simpler, being mainly to cause softening and the complete or partial elimination of cold work.

The situation in welding is seldom quite as simple as the position indicated in the above discussion because parent metals are often imperfect in detail and it is also possible for the molten pool to introduce hydrogen into the heat affected zone. The heat affected zone is, therefore, a site for potential defects and its behaviour in any given material is an important part of the consideration of weldability. Weldability, however, is a material property which cannot be defined precisely as it varies with the process employed and the way that process is used. Materials with indifferent weldability can often be welded satisfactorily but only as a result of extreme care in consumable selection, control during welding and inspection afterwards. This often means extensive preproduction trials and of course a cost penalty.

HAZ defects

Some of the defects which can occur in the HAZ are:

1. HAZ hydrogen cracking (sometimes called underbead cracking).
2. Lamellar tearing.
3. Reheat cracking.
4. Stress corrosion cracking.
5. Micro-cracking or liquation cracking.
6. Porosity.

HAZ hydrogen cracking

This type of cracking can occur in steels and results from the presence of hydrogen in a hardened microstructure suspectible to cracking, such as martensite, together with stress. Usually not much can be done about stress although it is known that joints with excessive gaps are more susceptible to cracking. The practical measures to prevent cracking therefore hinge on reducing the hydrogen in the molten pool and avoiding a hardened HAZ.

The previous section described how the molten pool can provide a source of hydrogen which diffuses in the austenite phase into the HAZ. When the region close to the weld transforms on cooling movement of hydrogen is slowed down, and it tends to remain where it can cause cracking (see Fig. 2.9). The hydrogen level is controlled by selecting a

Fig. 2.9 *Heat affected zone hydrogen cracks in a fillet weld made with a rutile electrode.*

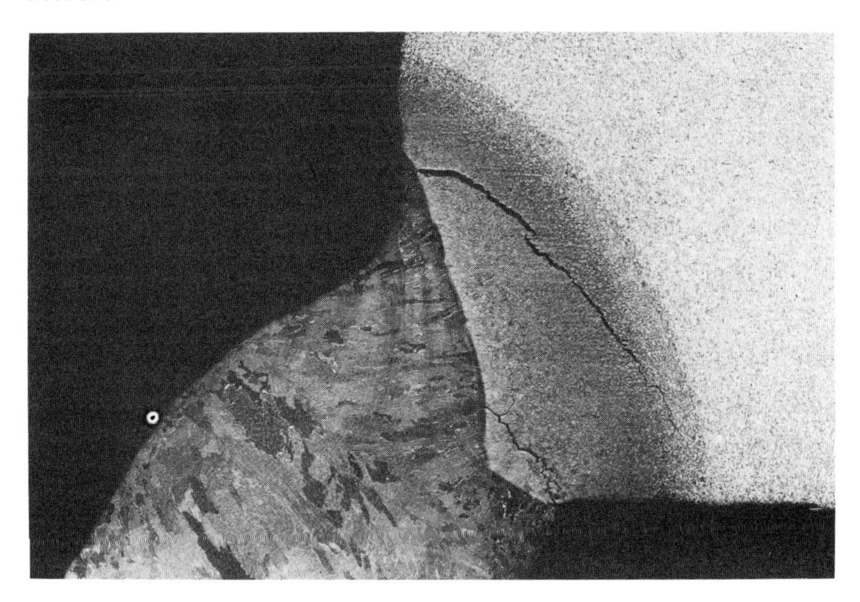

suitable type of welding consumable and by ensuring that it is dry. Rutile electrodes yield weld metal with a higher hydrogen content than basic electrodes and are preferred for the higher strength steels and those over about 25 mm (1 in.) thick. When welding highly sensitive steels, such as armour plate, an austenitic stainless steel electrode may be used as this weld metal does not transform and provides a sink for the hydrogen.

For any given steel the hardness developed in the HAZ depends on the cooling rate, the faster the cooling the higher the hardness and the more likely the structure is to crack. An important factor influencing cooling rate is the mass of material being welded, the thicker the joint the faster the cooling. The type of joint also affects the cooling rate through the number of paths along which heat may be conducted. With a butt joint there are two paths, however, with a 'T' joint there are three, so the same size weld in a 'T' joint cools faster (see Fig. 2.10).

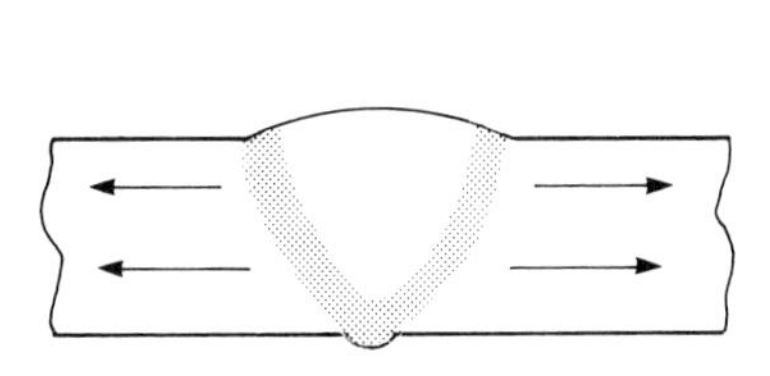

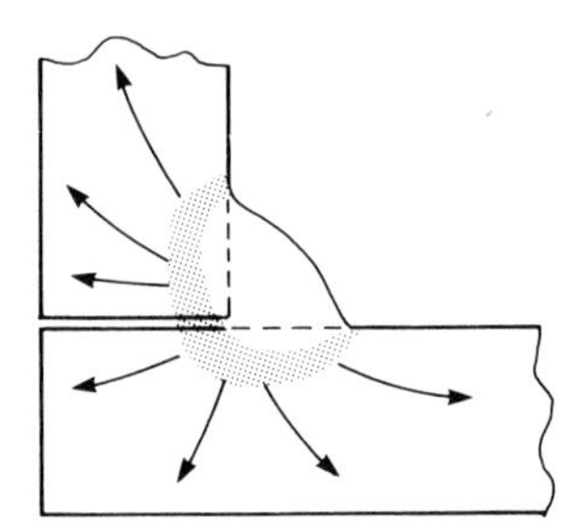

Fig. 2.10 *Heat flow paths in butt and fillet welds.*

Control of the microstructure is achieved mainly in two ways. Firstly, by choosing a steel which has a suitable hardenability. The hardenability of a steel is determined by its content of carbon and that of other alloying elements like manganese, chromium, molybdenum and vanadium, and various formulae have been devised for estimating carbon equivalent from the composition of the steel.

The carbon equivalent (CE) formula used in British Standards is as follows:

$$CE = C\% + \frac{Mn\%}{6} + \frac{(Cr\% + Mo\% + V\%)}{5} + \frac{(Ni\% + Cu\%)}{15}$$

NOTE: The CE calculated from 'ladle analysis' is usually lower than that based on the analysis of the plate or section because of the segregation which takes place in casting – see page 40.

Secondly, the microstructure may be controlled by slowing down the cooling rate which for any particular joint can be achieved in two ways: (*a*) through raising the heat input by increasing weld bead size and/or reducing welding speed. With metal arc welding this means using a larger diameter of electrode. Or, (*b*) by using a preheat. Hydrogen cracking occurs only at temperatures slightly above ambient so if the preheat is maintained after welding, for times depending upon the thickness of the steel, this helps the hydrogen to diffuse away harmlessly before cracking can take place.

A carbon equivalent less than 0.40 indicates that the steel is readily weldable but above this figure additional precautions with preheat or heat input may be necessary. As preheat is difficult and expensive to apply it is usually avoided if possible when manual metal-arc welding by employing basic rather than rutile type electrodes or in the extreme by using austenitic electrodes. Table 2.1 is a useful summary and further help in deciding the procedure to use can be obtained from the British Standard 5135: 1974 'Metal-arc welding of carbon-manganese steels'.

HAZ cracking is avoidable with care but is an insidious defect particularly in fillet welds where it may appear at the weld toe which is an area also likely to be a stress concentration. As a high cooling rate is a major contributory factor small welds like tacks (or even arc strikes) are potential sites for HAZ cracking and must be treated with as much respect as the main weld.

Table 2.1 *Summary of welding conditions for C-Mn steels.*

Carbon equivalent (%)						
up to 0.28	0.29–0.41	0.42–0.43	0.44–0.45	0.46–0.55	0.56–0.60	0.61–0.72
Butt welds and fillet welds up to 12 mm (0.47 in.)						
R/none	R/none	R/none	R/B 150	B/none (300 F)	B/100 (212 F) A/none	B/150 (300 F) A/none
Butt welds 15–21 mm (0.59–0.83 in.)						
R/none	R/100 (212 F) B/none	R/150 (300 F) B/none	B/none (300 F)	B/150	B/150 (300 F) A/none	B/250 (480 F) A/100 (212 F)
Butt welds 25–36 mm (1–1.4 in.) and fillet welds up to 18 mm (0.71 in.)						
R/none B/none	R/150 B/none	B/100	B/150	b/150	B/250 A/100	A only
Butt welds 37–50 mm (1.46–1.97 in.) and fillet welds up to 25 mm (1 in.)						
B/none	B/150 (300 F)	B/150 (300 F)	B/150 (300 F)	B/200 (390 F)	B/250 (480 F)	A only

The above table is for general guidance only and to indicate trends.
KEY: R = rutile electrode
B = basic electrode
A = austenitic electrode
Numbers following / indicate preheat in degrees C; those in brackets, degrees F

Lamellar tearing

This occurs in thick plate as a result of shortcomings in the plate material accentuated by welding strains and inappropriate joint design. Steel plate likely to be affected has poor through-thickness properties as a result of thin zones of non-metallic inclusions lying in layers parallel to the surface. These are opened up by welding strains to form stepped cracks close to the HAZ (see Fig. 2.11). The condition is aggravated by the presence of even small amounts of hydrogen. If it is suspected that a steel is likely to be susceptible to lamellar tearing, joints should be designed to avoid as far as possible weld contraction occurring in the through-thickness direction, e.g., by avoiding cruciform joints or heavy fillets, and low hydrogen well dried electrodes should be used. Buttering to protect the sensitive areas is helpful either before the main weld or by using in-situ buttering which is really a controlled sequence of welding (see Fig. 2.12). It is best, however, to estimate the risk of lamellar tearing before welding commences and if necessary to order steel plate with the appropriate through-thickness properties.

Reheat cracking

This may occur in some low-alloy steels at the grain boundaries, usually in the coarse-grained HAZ, after a weld has been in service at high

Fig. 2.11 *Lamellar tears in the heat affected zone of a multirun T butt weld.*

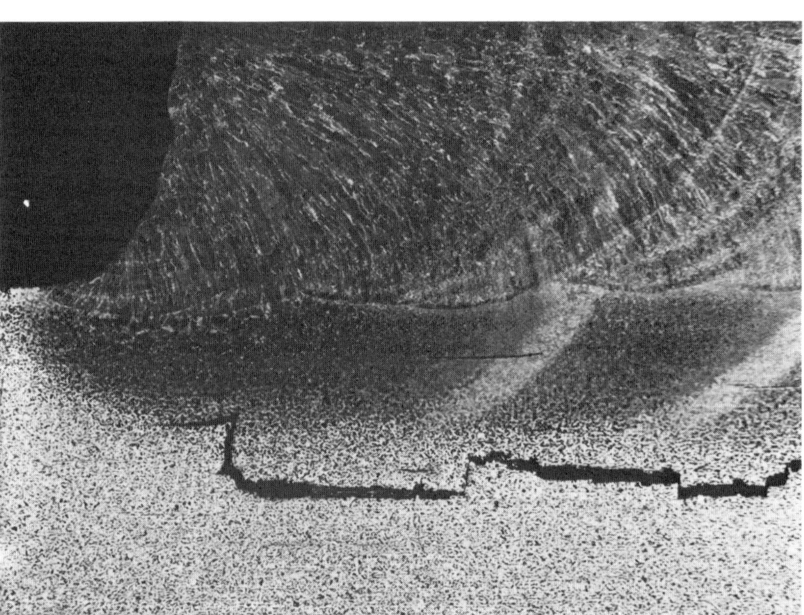

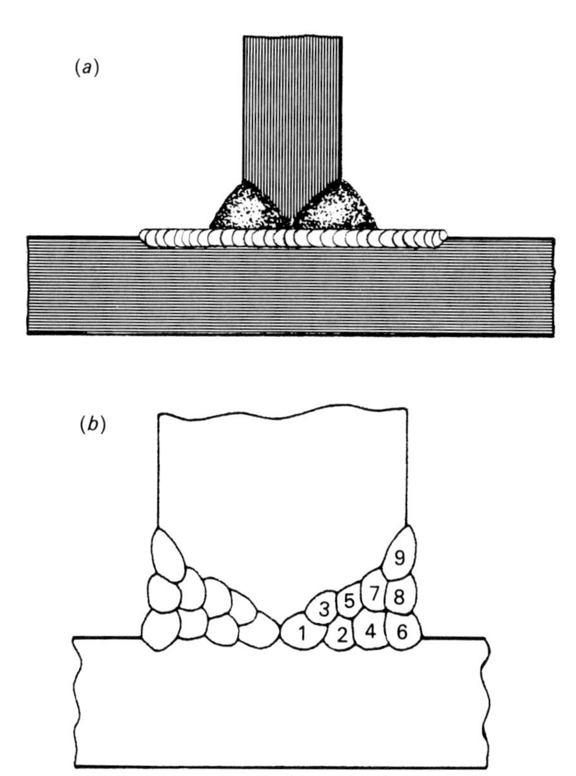

Fig. 2.12 *Principles of* (a) *buttering and* (b) *in-situ buttering to reduce risk of lamellar tearing.*

temperature or has been heat treated. The actual causes are complex and not completely understood but the mechanism may involve a stiffening of the interior of grains by carbide formers such as chromium, molybdenum and vanadium. This throws strain on the grain boundaries which if they are weakened by the presence of trace elements such as sulphur, phosphorus, tin, antimony and arsenic may then fail.

Intergranular stress corrosion

A form of cracking which can occur in many materials and is usually specific to a particular corrosive medium, e.g., hydrogen sulphide, can attack the hard region of the HAZ in line pipe steels. Hence a maximum hardness is frequently specified. General precautions against stress corrosion include careful selection of parent metal and a suitable post-weld treatment to reduce stress and put the HAZ into the most suitable microstructural condition.

Other possible defects in the HAZ include liquation cracking caused by the melting of low melting-point grain boundary constituents. This results in small (micro) cracks which may not be serious providing that they do not provide sites for more serious cracking. HAZ porosity which often takes the form of surface blistering can also occur in aluminium-magnesium alloys and tough pitch or other coppers containing oxygen. With the aluminium alloy this is because of hydrogen already present in the parent plate, but in the copper it results if hydrogen has been present in the weld metal and has diffused into the HAZ where it reacted with copper oxides to form steam.

Distortion

In all fusion-welding methods the weld is completed by moving a molten pool progressively along the line of the joint. Heat flows from the pool into the surrounding material where it affects the metallurgical structure (see later) and causes dimensional changes, some of which are permanent, resulting in distortion and residual stresses. At its simplest distortion is of two types: linear, caused by shrinkage in the length of the weld, and rotational, resulting from greater shrinkage on one face of the weld than on the obverse face. These two aspects of distortion are illustrated in Fig. 2.13 and Fig. 2.14. Of the two, rotational distortion is the more noticeable and can be minimised by making the weld bead symmetrical about the neutral axis or by having a parallel sided single pass weld, as with electron-beam welding. Overall distortion can be reduced by limiting the heat used to make the weld. The more intense the heat source and the faster the welding the less heat is conducted into the surrounding material so in any given joint the distortion is greatest when

Fig. 2.13 *Longitudinal distortion.*

Fig. 2.14 *Rotational distortion.*

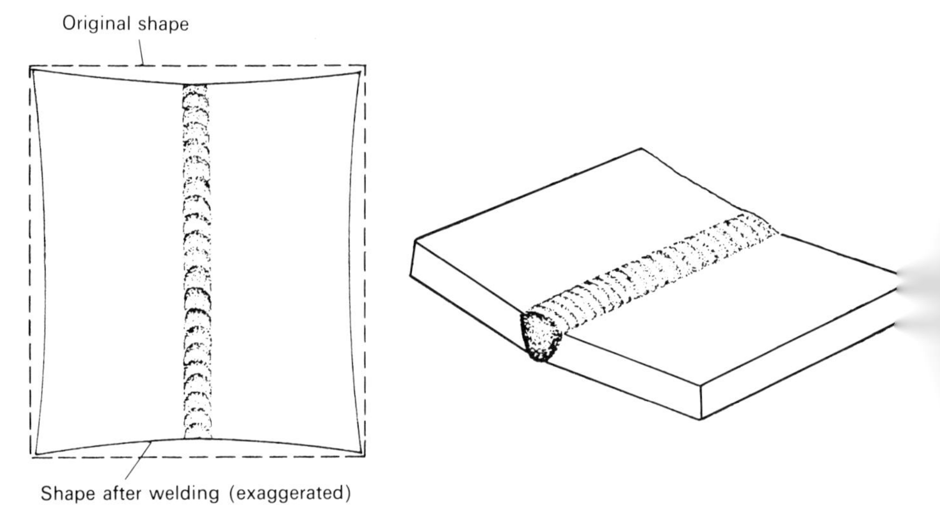

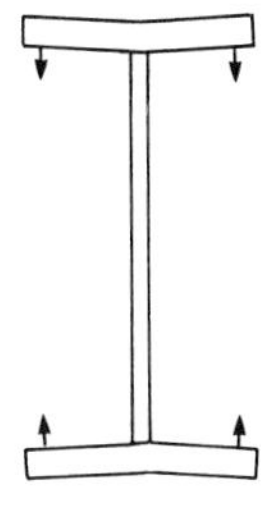

Fig. 2.15 *Allowing for distortion by pre-setting flanges.*

gas welding is used and least with the intense highly thermally efficient electron beam, with arc welding somewhere in between. In the workshop distortion is often minimised by offsetting the joints prior to welding (see Fig. 2.15) or by placing weld beads in a sequence, first on one side and then on the other side of the joint.

Weld metal requirements

Strength and toughness properties

Specifications frequently require weld metal to have certain mechanical properties or a particular chemical composition and the HAZ to have particular properties. With low-carbon mild steel it is not difficult to obtain a tensile strength in the weld metal which over-matches that of the parent plate thus ensuring that fracture in a tensile test takes place in the heat affected zone. Higher tensile steels require stronger weld metal which is obtained, not by increasing the weld metal carbon content, but by increasing manganese or making additions of molybdenum or nickel. With non-ferrous materials more careful specification of filler materials is necessary to ensure that failure in tensile loading takes place in the heat affected zone.

Transverse bend tests are frequently called for but are only a fairly crude way of showing up defects in the tension surface of the testpiece. The all-weld metal testpiece, a multipass pad of weld metal, is of value chiefly to establish the composition of the metal deposited from an electrode. It says nothing about the composition of a weld in an actual structure, however, as it takes no account of dilution. The mechanical properties measured on an all-weld metal testpiece allow comparison between consumables but cannot be correlated with actual joints since the metal is deposited under conditions which are not representative of real welds.

Weld metal in structural steels is frequently tested for toughness by the Charpy impact test which uses a 10 × 10 mm (0.394 × 0.394 in.) test bar notched in the centre and broken by impact. The more energy the

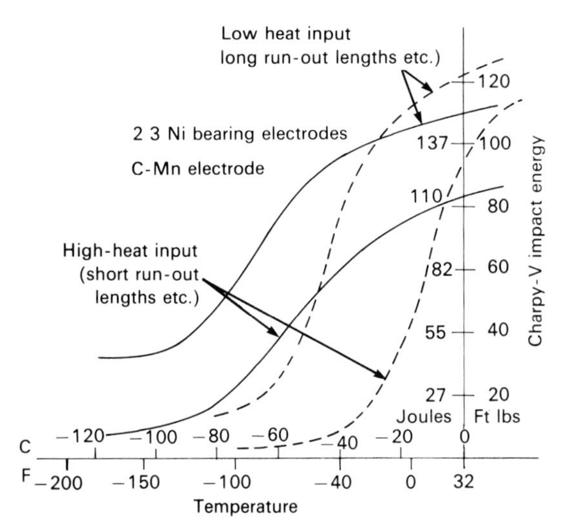

Fig. 2.16 *A charpy V notch transition curve.*

specimen absorbs during fracture the more ductile is the material. The test is carried out over a range of temperatures and the results are used on the basis of experience to indicate suitability of the weld for service, particularly at low temperatures or its susceptibility to brittle fracture. When the results of Charpy tests over a range of temperatures are plotted against temperature as in Fig. 2.16 a stepped curve is produced with many steels called the Transition Curve because it shows a transition in toughness as the temperature is changed. The high and low temperature levels are referred to as the upper and lower 'shelf' energies respectively. In the transition region itself there is frequently a considerable scatter in results.

An alternative test used for the same purpose but one which is amenable to fracture mechanics analysis is the Crack Tip Opening Displacement or CTOD test. The CTOD test, like the Charpy test also employs a notch but usually the machined notch is extended by a controlled fatigue crack. In this way the notch in the CTOD test is made to resemble as closely as possible the crack-like defects which occur from time to time in welds and which experience shows are frequently sites for brittle failure. A bend test in full section thickness is carried out with the notch in tension and the plastic deformation at the tip of the crack prior to failure is measured (see Fig. 2.17). As with the Charpy test the CTOD test is carried out at a reduced temperature appropriate to the intended service or over a range of temperatures.

Corrosion and heat resistance

Parent materials selected for their corrosion resistance must be welded

Fig. 2.17 *A COTD test in progress at ambient temperature. When testing at reduced temperatures the testpiece and lower support are immersed in a coolant.*

with the appropriate electrode or filler metal and usually under carefully controlled welding conditions. An exact match between weld metal and parent metal compositions is not always desirable and in practice with some materials certain stainless steels, aluminium alloys and copper alloys it is usual to raise the alloy content of the weld metal slightly above that in the parent metal.

In stainless steels, alloys of iron with at least 12 per cent chromium and 6 per cent nickel, corrosion protection is provided by a film of chromium oxide on the surface of the steel. Much investigation over several decades has produced a diagram based originally on one by Schaeffler (see Fig. 2.18) which indicates the various metallographic phases present for a range of chromium and nickel contents and their equivalents. This diagram indicates the regions which are susceptible to various kinds of defect and is a useful guide to the selection of parent metal and electrodes (see Chapter 3). Because the protective action of chromium depends upon it being in solution it is desirable to prevent it forming carbides. This is achieved either by having a very low carbon content or by adding a stabilising element such as titanium or niobium which combines preferentially with the carbon. Electrodes for welding stainless steels deposit low-carbon metal and some are stabilised but unlike the parent metal this must be done with niobium and not titanium which does not transfer well from the electrode to the molten pool.

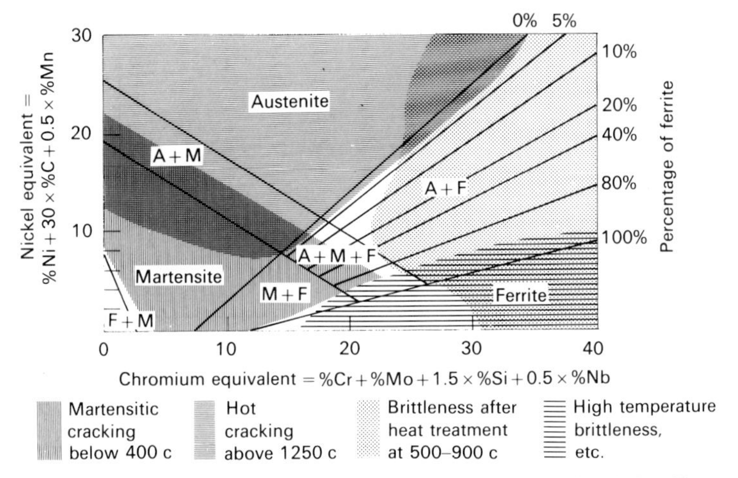

Fig. 2.18 *A constitutional diagram for stainless steel welds, after Schaeffler, with an indication of the weld defects which may occur in each constitutional region.*

For service at elevated temperature where oxidation resistance is important ferrous alloys high in both chromium and nickel are used. Elements forming stable carbides such as titanium, niobium, molybdenum, cobalt and tungsten are added. Creep resistance at temperatures up to 500 C (932 F) is a characteristic of steels containing chromium and molybdenum of which a group containing from 1–5 per cent Cr and 0.5–1.0 per cent Mo are commonly used in power plants. It is important that the weld metal in joints in these steels should also have appropriate creep resisting properties.

The composition and to some extent the thermal history of weld metal can be controlled by the welding engineer and operator to produce a joint of the required characteristics. There is less scope with the HAZ, where, with only a few exceptions, control is limited to thermal effects through the choice of process and the weld procedure.

Non-destructive testing of welds

Mention has been made of a number of defects which may be encountered in welds. These can be detected by a variety of well established methods for non-destructive examination – however, not all defects can be found with equal reliability by all methods. The principal methods are as follows:

Visual inspection

A method not to be overlooked! Good illumination is necessary and magnifying lenses are helpful. Special optical devices based on the telescope are available for inspecting inside vessels and other areas of limited access. Visual inspection will often reveal any surface-breaking cracks but is mainly important for showing poor weld shapes which may give a clue to faulty technique.

Dye-penetrant inspection

A liquid containing a dye or UV-sensitive substance is sprayed on the part to be examined and is drawn into any surface-breaking cracks, porosity or other defects. Its presence there is revealed after the application of a developing agent or by viewing under UV light. The method can only be used on clean surfaces.

Magnetic particle inspection

The welded part is magnetised locally by a yoke or coil, or by the passage of a heavy electric current. When magnetic powder particles suspended in a fluid are applied to the surface the magnetic discontinuities surrounding defects cause the particles to become concentrated near the defect. The method will detect surface-breaking and near-surface defects in ferrous materials.

X-radiography

X-rays are passed through the welded part and fall on a photographic plate placed in contact with its reverse side. In general, X-rays pass more readily through defects than through the surrounding metal and the defects therefore appear as dark areas on the developed film. Cracks, porosity, inclusions and lack of fusion are generally shown clearly. Tungsten inclusions appear light, not dark. Tightly closed cracks or lack of fusion aligned normally to the beam may not be detected. Gamma radiography uses isotope sources, e.g., Cobalt-60, instead of an X-ray tube. It is a portable method suitable for field use.

Ultrasonic flaw detection

Pulsed beams of ultrasound typically 1–25 MHz from a transmitter are passed through the welded part and are reflected from free surfaces and any weld defects. A receiver probe detects the reflected signals which after amplification and analysis are displayed on an oscilloscope or in various other ways. There must not be a free surface between the probe and the surface of the welded part, so contact is made through a film of oil. A single probe may act as both transmitter and receiver and inspection is usually from one side only. Ultrasonic flaw detectors are usually hand-held instruments but elaborate multi-probe systems or systems in which the welded part is immersed in a tank of water to be scanned mechanically are in use. The method is very sensitive and is capable of finding most weld defects and it can be used on welds of considerable thickness, greater than is normally possible with X-rays.

Chapter 3

Procedures for specific materials

As was explained in the first paragraphs of Chapter 2 a basic knowledge of the procedures to be adopted for the common materials is important to anyone using welding. The various codes and specifications issued by national and international bodies can be consulted at a number of libraries of which The Welding Institue holds one of the most comprehensive stocks. See section on Sources of Information and Bibliography, pages 224 and 226. The information which follows was prepared for manual metal-arc welding but many of the considerations apply equally well to most other welding processes.

Structural steels

The requirements for weldable structural steels appear in the British Standard 4360. These steels have been separated into four groups depending upon their minimum tensile strength, i.e., Grade 40, Grade 43, Grade 50 and Grade 55. Each grade is further divided into subgroups with differing chemical composition, yield stress, impact requirements, etc. The most important departure from earlier specifications is that a product analysis (i.e., the composition of the actual plate, section, etc.) is now quoted. Many of the steels in BS 4360 may be ordered to a maximum specified Carbon Equivalent (CE) which will be based on the ladle analysis. The actual CE in the plate or section may well be higher than this because of segregation allowed for in the product analysis. Thus, although a steel may be ordered to a maximum CE of say 0.41 on the ladle analysis, the CE of the actual product could well be up to 0.46.

There are three grades, Grade 40A, Grade 43A1 and Grade 43A, where no upper or lower limits are given for manganese and, therefore, carbon equivalents cannot be determined. The only limiting factors in these grades are the maximum carbon content and the upper tensile limit. If the carbon of the ladle analysis and the upper tensile limit on the Test Certificate of the steel supplied are both high then the manganese levels should be checked. This applies especially to thick sections and borderline cases in the choice of low hydrogen or rutile electrodes and whether to preheat or not. If in doubt as to the correct welding procedure reference should be made to BS 5135.

For USA equivalents, consult a reference book such as the ASM *Guide to equivalent irons and steel* (see page 226) or the AWS 'Structural Welding Code, D1.1', 1988.

Steels for low-temperature applications

High-quality basic coated low-hydrogen electrodes of the C-Mn type are generally suitable for welding materials with tensile strengths up to

Table 3.1 *Low temperature steel application guide.*

Type of steel	Typical specification		Recommended electrode type
C-Mn low temperature	BS 1501 LT 0	BS 4360 43A	Basic
C-Mn low temperature to −30 c (−22 F)	BS 1501 LT 30	BS 4300 50D	High quality basic
C-Mn low temperature to −50 C (−58 F)	BS 1501 LT 50	BS 4360 40E, 43E and 55E	2½% nickel
C-Mn and low alloy to −70 C (−94 F)	ASTM A333 grades 1, 6 and 7		BS 2493 2 Ni BH BS 2493 3 Ni BH
3% Ni low temperature to −101 C (−150 F)	ASTM A333 grades 3 and 4		3% nickel BS 2493 3 Ni BH AWS 8106/8108–C2
9% Ni low temperature to −196 C (−320 F)	ASTM A333 grade 8		Special nickel chrome alloy
Higher tensile low alloy (650 N/mm^2 (94.3 ksi) and above			Low alloy basic
	RQY 600 BS 4360 55FM API 5LX 65 and 70		AWS 8016
	QT28 QT35 HY80 NQ1		AWS 9018, 10018 11018, etc.
	LOYCON QT QT445 A and B		10018 11018, etc.
Supereiso 70			

about 650 N/mm² (94.3 ksi) and minimum charpy V notch impact properties at −40 C (−40 F) of 34 joules (see Table 3.1). Where there is a requirement to meet minimum charpy figures at −50 C (−58 F) it will be necessary to weld with either nickel-bearing electrodes or the latest C-Mn types developed as a result of innovations in electrode coating formulation. The manufacturers should be consulted for specific recommendations and the welding procedure will include the stipulation of minimum runout lengths, bead sequence and maximum welding current. Steels with various nickel contents are used for temperatures down to −196 C (−320 F), e.g., 3 per cent nickel steels are used for cryogenic applications down to −101 C (−150 F), whilst 9 per cent nickel steels are used for LNG applications operating at −196 C (−320 F). Electrodes of similar composition are used to weld steels containing up to 3 per cent nickel. Steels of the 5 per cent and 9 per cent nickel types are welded with specially alloyed nickel-chromium-iron electrodes, since electrodes containing above 3 per cent nickel are susceptible to solidification cracking.

Creep-resisting steels

These are used in power stations for the construction of steam lines and chests, turbine equipment, etc. and contain additions of chromium, molybdenum and sometimes vanadium to provide creep strength at operating temperatures up to 500 C (932 F). Basic coated low-alloy hydrogen controlled electrodes of similar chromium and molybdenum content are used for welding these materials. Two types of cracking may occur when welding these steels:

1. transverse cracking in the weld metal, and
2. heat affected zone cracking in the parent plate.

These steels must therefore be welded under strictly controlled conditions. This includes the use of preheating and interpass temperatures of 200–300 C (392–572 F) followed by a stress relief at approximately 700 C (1292 F), the use of smaller-size electrodes (usually 4.0 mm (5⁄32 in. maximum), and control of electrode angle, etc. (see Table 3.2).

High-tensile low-alloy steels

Materials of this type have a carbon content higher than mild steel together with additions of chromium, nickel, molybdenum and vanadium. Hard structures form on cooling and to avoid cracking preheating temperatures up to 300 C (572 F) may be required followed in some instances by post-heat treatment to restore properties. Generally the

Table 3.2 *Creep-resistant steels' application guide.*

Steel type	Recommended electrode BS 2439	AWS	Typical preheat and interpass temperature C	F
½ Mo	Mo BH	E7016/18–A1	150–200	300–392
1 Cr ½ Mo	1 Cr Mo BH	E8016–B2	200–250	392–480
2¼ Cr 1 Mo	2 Cr Mo BH	E9016/18–B3	250–300	480–570
5 Cr ½ Mo	5 Cr Mo BH	E502–16	250–300	480–570
½ Cr ½ Mo ¼ V	2 Cr Mo BH	E9016/18–B3	250–300	480–570

largest diameter electrodes possible should be used in combination with a welding technique balanced to minimise restraint. The use of austenitic stainless steels specially designed for the purpose, combined with preheat, may be necessary to prevent cracking in the more highly alloyed material as typically used for transmission parts, gears, some armour plate, etc.

Stainless and heat-resisting steels

Stainless steel and heat-resisting steels are a group of iron-based alloys containing at least 12 per cent chromium which confers corrosion and heat resisting properties. Nickel in amounts of 6 per cent and more improves corrosion and heat resistance and also makes fabrication of these materials much easier. For very high temperature service a nickel content of 15 per cent and over is used to maintain strength. Other alloying elements such as molybdenum, titanium, tungsten, niobium, copper and cobalt are sometimes added in relatively small amounts to improve corrosion resistance or high temperature strength or creep resistance.

A useful method of assessing the general metallurgical characteristics of any stainless steel weld metal is by means of Schaeffler's diagram (see Fig. 2.18). The various alloying elements are expressed in terms of nickel or chromium equivalents, i.e., elements which like nickel tend to form austenite and elements like chromium which tend to form ferrite. By plotting the total values for the nickel and chromium equivalents on the Schaeffler diagram a point can be found indicating the main phases present in the stainless steel and this provides information as to its behaviour during welding.

The diagram indicates that the comparatively low alloyed steels are hardenable since they contain the martensitic phase in the as-welded state. As the alloying elements increase, the austenite and ferrite phases become more stable and the alloy ceases to harden on quenching. Steels with a relatively high level of carbon, nickel and manganese become fully austenitic ('Austenite' area), while those with more chromium, molybdenum, etc., tend to become fully ferritic ('Ferrite' area). There is also an important intermediate region of 'duplex' compositions indicated as A + F on the diagram. In this region the welds contain both austenite and ferrite. This leads to the general classification of stainless steels into austenitic, ferritic and martensitic, according to which phase is predominant. Superimposed on Schaeffler's diagram are shaded areas which indicate approximate composition regions in which defects may, under certain circumstances, appear in stainless steel welds.

Martensitic region

The martensite area comprises low-alloy hardenable and martensitic stainless steels and is associated with a hard and brittle structure formed as a result of cooling from high temperatures. Weld metals of the 13 per cent chromium type fall into this group. To counteract the hardening effect of cooling (both in the weld and the heat affected zone of the plate), and the associated risk of martensitic cracking, it is advisable to slow the rate of cooling by pre-heating. A post-weld heat-treatment may also be desirable to soften the heat affected zone. Steels of this type are used for cutlery, spindles, shafts and applications requiring good resistance to corrosion and scaling at elevated temperatures up to approximately 800 C (1475 F). For ductile joints in 12–16 per cent Cr alloy free from cracks, an austenitic stainless steel electrode is normally used, electrodes depositing weld metal of similar composition usually being limited to applications such as overlaying and welding minor attachments, or when it is important to match corrosion/mechanical properties.

Austenitic region

The fully austenitic region comprises compositions such as 18Cr–13Ni, and 25Cr–20Ni. Though fully austenitic compositions offer certain advantages in forming and in resistance to corrosion and creep, they display a tendency to solidification cracking and fissuring during welding under conditions of severe restraint. The composition areas likely to experience hot-cracking are indicated in Fig. 2.18. The 'duplex' region (A + F) includes the most popular austenitic steels. There is a minimum of 18 per cent chromium and 8 per cent nickel in these steels but there are also many special alloys with a greater percentage of both elements such as 25/20, 18/12, etc., and with smaller additions of other elements to provide improved corrosion resistance under specific conditions. For example, the presence of 2–3 per cent molybdenum increases resistance to sulphuric acid. The term 'austenitic' implies, from the practical point of view, that steels of this composition cannot be hardened by heat-treatment. In fact drastic quenching from a high temperature actually softens them. Mostly non-magnetic, these steels harden rapidly when cold worked, though initially they are nearly as soft as ordinary mild steel and are capable of as much cold working if frequently annealed. Austenitic stainless steels are used for a great variety of purposes, e.g., the chemical and food industries, furnace parts, gas turbines, etc.

The technique of welding austenitic stainless steel does not differ greatly from that for mild steel but as the material is expensive extra precautions and attention to detail at all stages of fabrication are desirable, e.g., edge preparations must be clean with good fit-up. Preheat and post-heat treatments are not normally required. The tendency to distort is much greater than with mild steel, so jigs are used wherever possible, particularly for material thinner than 3.2 mm (1/8 in.), and tack welds should be made at intervals of about half the pitch used for mild steel. Weld deposits should be balanced by a step-back technique or similar sequence to minimise distortion. The lowest convenient current should be used with a weave no greater than twice the diameter of the electrode. On thin sheet it is an advantage to increase the speed of travel by welding in the semi-vertical down position. This keeps heat input low, helps to prevent burn-through and distortion and gives a neat weld profile. A short arc should be held to avoid loss of alloy elements across the arc.

When welding austenitic stainless steels an electrode of matching composition to the base material is normally used. Most electrode compositions fall within this group. The welds contain residual ferrite and are resistant to hot-cracking in restrained joints. However, for certain applications in the temperature range of 450–900 C (840–1650 F) the grades with higher ferrite contents may suffer a gradual loss of ductility and impact resistance due to the transformation of ferrite to the brittle phase sigma. Fortunately most stainless steel is not subjected to conditions such as these and sigma phase embrittlement does not occur.

Ferritic region

A typical alloy within this area contains 18 per cent Cr. The good resistance to scaling and to sulphur-polluted atmosphere makes these steels useful for applications in furnace equipment. The welds made with the ferritic stainless steel of similar composition to the parent metal are subject to embrittlement due to grain coarsening when reheated above 1050 C (1922 F). In addition, the impact resistance is inherently low because of high brittle transition temperatures. Although these steels are tough at red heat they are brittle at normal temperatures. If grain coarsening occurs this will accentuate the brittleness. Matching consumables are generally used. Ferritic steels are frequently used in sulphur-bearing atmospheres which attack nickel but in mildly corrosive applications and where the presence of nickel-bearing weld metal can be tolerated an austenitic stainless steel electrode is recommended. A weld of this type also provides a joint capable of deformation in further processing operations.

Duplex stainless steels

Stainless steels of the 'duplex' type have compositions in the range 18–26 per cent Cr, 4–8 pr cent Ni, 0–3 per cent Mo and sometimes other elements such as copper and nitrogen. The steels will normally have received a solution heat treatment at around 1050 C (1922 F) giving a uniform two-phase structure of ferrite and austenite. Heating and subsequent reheating in multipass welds cause the alloys to transform to ferrite and then to revert to austenite on cooling. On heating, carbides and nitrides are taken into solution to be re-precipitated on cooling or re-heating. Unlike the austenitic steels, duplex steels are not stabilised against intercrystalline attack by titanium or niobium and the precipitation referred to above is mainly chromium carbide or nitride. With the duplex steels resistance to intercrystalline attack is conferred by the low carbon level of the alloys, i.e., less then 0.03 per cent. The alloys are in fact considerably more resistant to stress corrosion cracking in chloride-containing media than the austenitic grades and are commonly used in plant operating in the range 0–300 C (32–572 F). They are also stronger alloys and therefore have some structural advantage.

Schaeffler's diagram can generally be applied to duplex steels but it should be noted that nitrogen is a powerful austenitising element with almost the same level of effect as carbon. The proportion of austenite in the weld metal and HAZ is influenced by both composition and thermal history in the weld but mainly by composition. Dilution particularly in TIG welding must be considered. To obtain both satisfactory corrosion resistance and mechanical toughness at least 30 per cent austenite is usually required in weld and HAZ.

Duplex alloys are readily welded by all processes but to ensure the optimum level of austenite it is usual to employ consumables higher in nickel and/or nitrogen than the parent metal. The alloys are sensitive to embrittlement following prolonged exposure to temperatures over 300 C (572 F) and therefore a maximum interpass temperature of 200 C (390 F) is recommended.

Welding of stainless steels to mild and low-alloy steels

Situations frequently arise when it becomes necessary to weld an austenitic stainless steel to a mild or low-alloy ferrite steel. In selecting a suitable electrode, the effect of dilution of the weld metal by the parent material must be considered. Dilution may be from 20–30 per cent depending on the welding technique used, root runs in butt joints being the most greatly affected. If a mild or low-alloy steel electrode is used to weld stainless to mild steel, the pick up of chromium and nickel from the stainless steel side of the joint could enrich the weld metal by up to 5 per cent chromium and 4 per cent nickel. This would result in a hardenable crack-sensitive weld. Austenitic stainless steel electrodes are therefore used for joining dissimilar metal combinations of stainless materials to mild and low-alloy ferritic steels. Sufficient alloying elements to overcome the effects of dilution from the mild or low-alloy steel side of the joint are necessary as weld metal which does not start with an adequate alloy content may give a final weld containing less than 17 per cent chromium and 7 per cent nickel. Weld metal with lower chromium and nickel contents are crack sensitive. Also, if as a result of dilution the weld metal is incorrectly balanced with nickel and chromium, there may not be sufficient ferrite present in the weld metal to prevent fissuring and subsequent cracking taking place. For these reasons the austenitic stainless steel electrodes of the 20/9/3 or 23/12 type whose composition has been specially balanced to ensure that the total alloy content is adequate to accommodate dilution effects and which have a ferrite content sufficient to provide high resistance to hot cracking should be used. Electrodes of the 25/20 type are also suitable for most applications except for conditions of high weld restraint.

A further advantage of using austenitic filler material to weld alloy steels is that the risk of HAZ hydrogen cracking is reduced. Unlike ferritic weld metal austenitic deposits retain their capacity to dissolve hydrogen so that diffusion into the HAZ is retarded.

Welding of clad steels

The use of a clad-material, consisting of a mild or low-alloy steel backing faced with stainless steel, usually from 10 to 20 per cent of the total thickness, combines the mechanical properties of an economic backing material with the corrosion resistance of the more expensive stainless facing. This facing usually consists of austenitic stainless steel of the 18/8 or 18/10 type, with or without additions of molybdenum, titanium and niobium or a martensitic stainless steel of the 13 per cent chromium type. The backing should be welded first, at the same time making sure that the root run of the mild steel electrode does not come into contact with the alloyed cladding. This can be achieved in two ways either by cutting the cladding away from both sides of the root, or

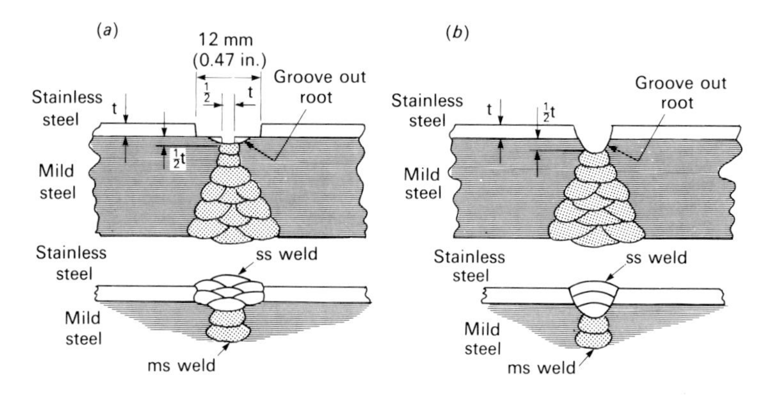

Fig. 3.1 *Techniques for welding clad steel:* (a) *cladding cut away from both sides of root;* (b) *root face left, followed by back-gouging and then welding with stainless steel.*

welding with a close butt preparation and a sufficiently large root-face (see Fig. 3.1).

After welding the mild steel side the root run should be back grooved and the stainless clad side welded with a stainless electrode of matching composition. The use of a more highly alloyed electrode (e.g. 25/20 type) for the initial root run on the clad side is advisable. This applies particularly to preparations in which the back-cutting of the cladding makes pick-up from the mild steel difficult to avoid. For the best resistance to corrosion, at least two layers of stainless weld metal on the clad side are recommended.

The welding of material which is clad or lined with 13 per cent Cr (martensitic) steels usually requires a preheat of 250 C (480 F) and the use of austenitic electrodes of appropriate type. Welding should be followed by a post-heat treatment, though satisfactory results can be obtained without these precautions if, during welding, heat dissipation is kept to a minimum. This will help to temper the heat affected zone by utilising the heat build-up from adjacent weld runs.

Hardfacing

Wear is a natural consequence of all mechanical plant operation. As loadings and speeds increase, the rate of wear is accelerated and replacement costs can become severe. It is then that repair or reclamation by welding can be an important contribution to overall economic operation. Wear may be caused by impact, abrasion, corrosion and heat or any combination of these factors. For practical purposes the type of wear most generally encountered may be considered as:

Shock (or impact) Material to resist this type of wear must be sufficiently hard to resist serious deformation but not so hard as to be brittle and to crack under the effect of shock or impact loading.

Abrasion This type of wear is caused by the grinding action of non-metallic particles against the surface. To resist this type of wear a relatively hard material is required but it often happens that a material of this type is also brittle and unable to withstand severe impact without cracking.

It is seldom that either of these types of wear occur alone, both being present to a greater or lesser degree, so the selection of a consumable that will cope satisfactorily with both conditions is a compromise (Table 3.3).

When building up large areas it is advisable to divide the work into small sections and to weld in alternating positions so that heat and contraction stresses are not cumulative. As a general rule it is advisable to avoid thick deposits of hardfacing alloys, since they are unable to accommodate the considerable shrinkage stresses set up and spalling of the deposit is liable to occur. To minimise the effects of dilution of the weld a buffer layer is essential when welding on to materials high in carbon or alloying additions. For this purpose a nickel-based electrode or in special cases one of the austenitic stainless steels are used on cast iron, while a basic mild steel electrode is used on high tensile steels. When a heavy deposit has to be built up with multiple runs of weld metal, or where service conditions impose impact loading on the part

Table 3.3 *Selection of consumables for surfacing.*

Condition required	Type of deposit	Approx. hardness
High impact resistance with medium abrasion resistance	Low alloy	350 HV
Medium impact and high abrasion resistance	Medium alloy	650 HV
Excellent impact resistance and work-hardening under impact to resist abrasion	13% Manganese steel or Austenitic steel	250 HV (500 HV when work hardened)
Maximum resistance to impact and abrasion	Chromium carbine	700 HV
Maximum abrasion resistance with moderate impact resistance	Tungsten carbide	1800 HV

NOTE. Hardness figures quoted are approximate and will depend upon parent metal and deposition procedure.

under repair, a buffer layer between the hardfacing and the underlying metal is also needed. The electrode chosen for this buffer should give a hardness value somewhere between that of the underlying metal and that of the final deposit.

Preheating may be required to do the following:

1. To prevent cracking when the material being surfaced has a sufficient carbon or alloy content to make it hardenable.
2. To prevent cracking of rigid brittle components as a result of thermal shock or contraction stresses set up by the deposited weld. Materials which are susceptible to cracking from this cause are high tensile alloy steels, high carbon and tool steels, and cast irons. As a general rule cracking in brittle materials may be prevented by slow heating to 400–600 C (750–1100 F) followed by very slow cooling.
3. To prevent the cracking of large areas of the particularly hard types of deposit.
4. To minimise distortion.

Note that 13 per cent manganese steel is never preheated, see below.

The 13 per cent manganese steel behaves in some respects in an opposite manner to ordinary carbon steel. When quenched from 1000 C (1830 F) it is austenitic and both soft and tough but when attempts are made to cut or abrade it rapid work hardening sets in producing extreme hardness. If on the other hand it is cooled slowly the structure becomes martensitic and is hard and brittle. The performance in wear can be seriously reduced if the material is excessively heated during welding, the degree of embrittlement which occurs being greater as the temperature and heating period is increased. Preheating or stress relieving should never be used and the minimum current should be employed. Dilution should be avoided and heat build-up minimised, e.g., by staggered welding so that the temperature is below 200 C (390 F). Any surfaces prepared by thermal cutting should be ground prior to welding.

Welding cast irons

Cast iron is a general term which describes material of widely varying compositions and physical characteristics, which may be divided into the following groups:

Grey iron This is so-called because of the grey appearance of the fracture. This material contains between 2.5 and 5 per cent carbon, mainly in the form of flake graphite, and up to 3 per cent silicon. It is made by slow cooling of the casting, and is relatively soft and weak.

White iron This is of similar composition to grey iron but with a lower silicon content. The carbon is present in the form of hard and brittle cementite and martensite as a result of rapidly cooling the casting with 'chills'. The name is derived from the characteristic white appearance of the fracture. Because of its extensive brittleness, white iron is unweldable.

Malleable irons (white heart and black heart) These are white cast irons which have been heat-treated to render them more ductile than grey irons.

Alloy irons These are made for wear, corrosion and heat-resistance, and for extra strength. Examples are 'Ni-Resist' (corrosion resistance), 'Nicrosilal' (heat resistance) and 'Meehanite' (high tensile). Some of these cast irons contain sufficient alloying elements to make them austenitic.

Spheroidal graphite iron (SG iron, ductile cast iron, nodular cast iron) By the addition of a small amount of magnesium (generally as nickel-magnesium alloy) during tapping into the ladle the graphite is made to form spheroids uniformly dispersed throughout the structure instead of the usual flake form. The result is a cast iron which, in the annealed state, has mechanical properties similar to those of mild steel. It has considerable ductility and shock resistance and is easier to weld than other types of cast iron. However, martensitic hardening of the heat affected zone may still take place and proper precautions must be taken during welding if this is to be avoided.

Welding of these materials is usually for the purpose of repair or salvage. Preheating, preferably of the whole component to be welded, to a temperature over 450 C (840 F) is desirable, followed by slow cooling. This reduces the chance of cracking as a result of cooling stresses and softens the HAZ which would otherwise harden because of the rapid heating and cooling during welding. The weld metal will inevitably pick up carbon from the parent metal by dilution and where an ordinary mild steel electrode is to be used this would create a hard brittle weld. This effect can sometimes be obviated by preheating and slow cooling but generally an electrode such as nickel or nickel-copper which is not sensitive to carbon pick-up is used. The deposits from these electrodes are soft enough to be machined easily. When a large multi-run weld is required the nickel or Monel electrodes may be used to produce a 'buttered' layer on the cast iron to enable basic covered low hydrogen mild steel electrodes to be used to fill up the joint. For welding alloy irons a nickel iron alloy electrode is particularly suitable as it provides a stronger deposit.

Welding non-ferrous metals

The alloys of aluminium and copper are difficult to weld because of high thermal conductivity, high thermal expansion and a tendency to weld cracking and porosity. Manual metal arc welding is possible on a few of the alloys but greater success is usually achieved with the gas-shielded arc processes, gas tungsten arc or gas metal-arc.

Chapter 4

The manual metal arc process

Frequently used names for the process – MMA, manual metal arc; Stick electrode welding; SMA, shielded metal arc

The historical introduction has given an indication of the origins and characteristics of the manual metal arc process but did not emphasise the vital part played by this relatively simple process in the general development of welding. Until the early 1980s more weld metal was deposited by this process than by all the other processes put together, and it was the manual metal arc process which established welding as the dominant method of joining metals and made the uptake of the newer methods which came later so much easier.

The simplicity of the process is illustrated by Fig. 4.1 and Fig. 4.2, but the flux-covered electrode itself is far from simple and has been, and still is, subject to continuing development. Much is now known about the function of the different minerals and compounds which make up the covering which is extruded on to the steel rod. Inspection of the

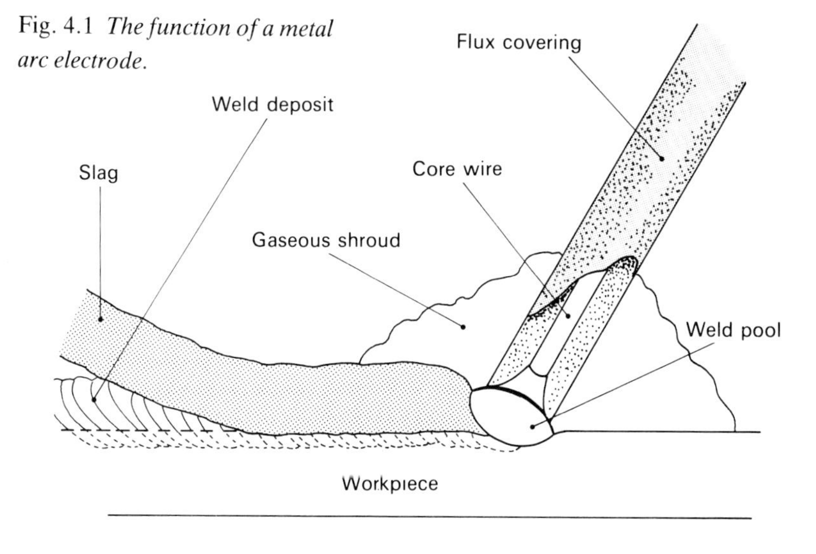

Fig. 4.1 *The function of a metal arc electrode.*

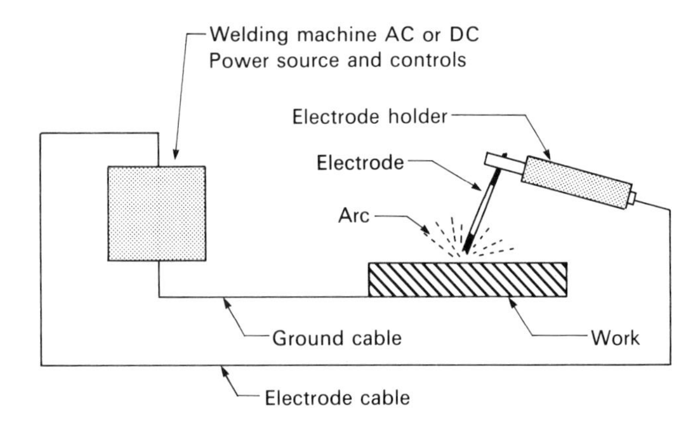

Fig. 4.2 *Circuit diagram for the MMA process.*

catalogues of consumable manufacturers will show that coverings can be compounded to make the electrode suitable for a wide variety of applications, different metals and welding positions. Welding electrodes are identified by the diameter of the core wire which normally ranges from 1.6–6.3 mm (1/16–1/4 in.) but on occasions electrodes of 10 mm (0.39 in.) or more in diameter are used. The length of individual electrodes usually varies from 250 mm (10 in.) for the smallest diameters to 450 mm (18 in.) for the largest, but for special purposes, e.g. gravity welding (see later) longer electrodes may be required. At one end of the electrode the flux is removed during manufacture to provide a gripping and current contact point for the electrode holder attached to the welding cable (see Fig. 4.3).

Once the arc has been struck, by a motion like the striking of a match, the electrode will begin to melt and a downward motion of the electrode holder is required to keep a constant arc length. When all but about 50 mm (2 in.) of electrode has been consumed welding must stop to allow the stub end to be removed and a new electrode to be inserted in the holder. While this is being done the weld bead freezes forming a crater. At this point it is usual to remove the slag by chipping and to begin the new deposit by moving the electrode momentarily back up the crater away from the direction of welding. This ensures that cracks or porosity in the crater are melted out. This intermittent operation is inevitable with the manual metal arc process and while it has disadvantages, e.g., weld defects can occur at the electrode change-over points, it is not possible for a welder to weld continuously in any case. Many constructional jobs require welds to be made in a number of different locations remote from the source of power. This type of application is ideal for the manual metal-arc process, the equipment is

Fig. 4.3 *The manual metal arc process in use.*

light and can be attached to a cable as long as 30 metres (100 ft) or more. In fact the limitation on where the process can be used is sometimes decided not by the process but by whether or not the welder can reach the weld position. The manual metal arc process is known by the abbreviation MMA in the UK but as SMA in the US.

Principles

When the arc is struck from the end of the electrode the core wire begins melting and melts away faster than the flux covering which surrounds it. The result is a cup shaped end to the electrode, the depth of which depends on the nature and thickness of the flux covering. Metal is transferred from the core wire as a spray of drops when the covering contains much rutile or in larger drops which touch the pool into which they become incorporated, when the covering is mainly basic constituents such as calcium carbonate. The transferring drops and the weld pool are partially shielded by a film of molten flux and partly by gas, mainly carbon dioxide but also hydrogen in the case of a cellulosic covering. These gases are generated by chemical reactions of the constituents of the flux covering in the heat of the arc. In addition to its shielding role the flux covering has to ensure that the deposited metal has the appropriate chemical composition and through this the desired mechanical properties. Electrode coverings therefore contain deoxidants the most common of which for ferritic steel electrodes is silicon. This is introduced into the covering as ferro-silicon. Additions of ferro-manganese or manganese metal are also made. It would be technically feasible to make some of these additions through the core wire but this is rarely done with ferritic electrodes for welding mild and carbon manganese steels for which a low carbon steel wire is employed. The flexibility of the metal arc process owes much to the ease with which the composition of the flux can be adjusted to suit individual needs.

The fused flux residue, the slag, covers the completed weld bead. This flux cover is not just something left over from the shielding and deoxidation role, it has an important function in shaping the final surface of the weld bead. A fluid slag which is usually high in oxides results in a more fluid weld pool which 'wets' smoothly to the parent plate. The slag from basic electrodes selected for their ability to produce low-hydrogen, high-strength weld metal provides a less fluid slag which results in a more convex bead. For welding in position, that is other than the flat position where weld beads form under the action of gravity, a voluminous and slightly viscous slag helps to hold the weld metal where it is required. In high-speed vertical-down welding as employed for welding pipelines too much slag can interfere with welding by running ahead of the pool and an electrode producing minimal slag such as the cellulosic electrode is often used. When welding in deep grooves with many weld passes it is important tht the slag is easily removed otherwise the quality of the weld will be affected with slag inclusions and possible associated lack of fusion.

Equipment

Power sources

A great variety of power sources exists for use by MMA. Single-phase AC transformers are the cheapest power sources. Three-phase input provides a more even load on the mains and is always used for high-capacity high-current power sources. These high-capacity units may also be oil immersed for cooling although units with fan cooling or natural convection cooling are common. An open-circuit voltage of at least 60 V is desirable to ensure that as many electrode types as possible can be employed, although some electrodes work well at voltages considerably below 60 V. After the arc has been struck the welding voltage of 20–35 volts is relatively safe but when the power source is on open-circuit the voltage may be sufficient in certain operating conditions, e.g., when surrounding conditions are wet, to cause accidents because of electric shock. For extra safety in these conditions power sources can be fitted with an electrical device (Voltage Reduction Device – VRD) which reduces the voltage at the terminals when there is no arc. Attempting to strike the arc automatically makes the full voltage available at the terminals. Regulations on allowable open circuit voltage vary locally and should be checked before specifying the use of AC transformers.

Remote control devices are sometimes fitted to power sources to enable a welder to change current settings without having to leave his working position. This can be of some economic advantage when working some distance from the power source or in an inconvenient position. Such devices are motor-driven attachments to the transformer handwheel or remote controlled switchgear.

MMA welding with AC transformers is the cheapest option as far as equipment cost is concerned and the plant is remarkably robust. Most, though not all electrodes for steel work on AC and many actually provide easier welding than on DC. This is true for the high efficiency iron powder electrodes especially in large diameters. The major advantage of AC is the absence of arc blow which can be encountered when using DC. Only AC can be used for the gravity welding process (see later), however, many non-ferrous electrodes and certain highly alloyed steel types will not operate on AC.

DC for MMA can be supplied by transformer–rectifier or converters (motor-generator sets) but at considerably greater capital cost. For site welding the DC engine-driven generator is often used (see Fig. 4.4). Engine units are two-stroke petrol for the smaller capacity power sources, e.g., up to 170 amp output, and either four-stroke petrol or diesel units above this. Generally, electrodes designed for use on DC will operate on lower open-circuit voltages than AC electrodes. DC is essential for non-ferrous electrodes like aluminium alloys and copper alloys. Remote control of current is easier and more common with DC power sources than AC. With motor-generator sets the variable resistor controlling current can be removed from the power source and then connected to it through a cable. Magnetic amplifier sets already have remote control through the variable resistor used to control the saturating current. Electronic power sources are particularly easy to control remotely because the control circuits carry low currents and can be separated by lightweight cables from the main power source.

Fig. 4.4 *An engine-driven welding generator.*

Accessories for MMA welding

There is a wide range of accessories which a welder requires for either his comfort or safety during welding including:

Helmets These are necessary for the protection of eyes and head and incorporate filters to screen out ultra-violet and infra-red radiation and reduce the brightness of the arc. Filters are made in a range of densities to BS 679: 1979 'Filters for use during welding and similar industrial operations'. The higher the welding current the darker the shade. As a protection from damage by spatter a plain cover glass is often fitted in front of the filter.

Electrode holders These must be light, easy to hold and grip the electrode securely with the lowest possible electrical resistance. With the passage of several hundred amps even a small resistance can generate considerable heat. Holders should be insulated so that they may be laid on the workpiece without short circuiting the power source.

Welding return clamps (also called earth return clamps, or work return clamps) These must make secure and low-resistance contact with the workpiece and are generally made from a copper-based alloy.

Welding cables These connect the electrode and the workpiece to the power source and should be of a sufficient size not to overheat. They should be kept as short as possible and any connectors should also offer the minimum electrical resistance.

Chipping hammers and wire brushes These items are necessary to remove the slag at the end of a run and clean up the surface.

Various items of protective clothing Special clothing is necessary to give protection against heat and hot spatter while safety spectacles are used when chipping off slag.

MMA consumables

Electrode coverings for MMA, specifications for which are shown in Table 4.1, are of three main types, cellulosic, rutile and basic, each of which may be modified by the addition of iron powder. The table summarises the specifications of AWS, ASTM and BSI for mild steel electrodes. Specifications are also issued by other national bodies.

Formerly, acid electrodes were also much used in addition to the types mentioned above. They had a high proportion of iron oxide in the covering and gave smooth well-shaped weld beads tending to be concave in fillets but of relatively low strength. The slag was voluminous and easy to remove. These electrodes have now been replaced by rutile electrodes which are easier to strike. Acid electrode coverings containing iron powder are, however, used for welding in the flat position (sometimes called 'downhand' position) and for gravity welding but only on the lower strength mild steels. See Fig. 4.5 for welding positions.

Iron powder additions

Additions of iron powder to a flux covering have two important effects, (*a*) to raise the rate of metal deposition, and (*b*) to modify the operating characteristics of the electrode giving smoother welding, less spatter and a higher operating current for the same size of core wire. Iron powder present in an electrode covering is fused during welding and finds its way into the weld bead so increasing the weight of metal deposited. Without iron powder the metal deposited is only 80–95 per cent of that in the core wire because of losses as spatter and fine particles. The metal recovery or electrode efficiency in this case is said to be 80–95 per cent. If iron powder is added to the covering then, depending on the amount, the metal recovery can be increased to 170 per cent or more. The higher recovery together with the opportunity to use a higher welding current

Table 4.1 *Summary of specifications for electrodes.*

BS 639
This specification is in two parts, one compulsory, the other optional.
Example E 51 3 2 RR (compulsory) 170 3 1 (optional)

Compulsory part

E denotes a covered electrode for manual metal arc welding

51 denotes tensile and yield strength as below:

43 indicates 430–550 N/mm^2 (62.3–79.8 ksi) tensile strength
330 N/mm^2 (47.8 ksi) yield strength
51 indicates 510–650 N/mm^2 (74–94.3 ksi) tensile strength
360 N/mm^2 (52.2 ksi) yield strength

3 denotes first digit for elongation and impact strength as below:

	Min. elong. %		*Temp. for impact value of 28 Joules*
	E43	E51	
0 indicates	Not specified		
1 indicates	20	18	+20 C (68 F)
2 indicates	22	18	0 C (32 F)
3 indicates	24	20	−20 C (−4 F)
4 indicates	24	20	−30 C (−22 F)
5 indicates	24	20	−40 C (−40 F)

2 denotes second digit for elongation and impact strength as below:

	Min. elong. %		*Impact Joules*		*Temp.*
	E43	E51	E43	E51	
0 indicates	Not specified				
1 indicates	22	22	47	47	+20 C (68 F)
2 indicates	22	22	47	47	0 C (32 F)
3 indicates	22	22	47	47	−20 C (−4 F)
4 indicates	§	18	§	41	−30 C (−22 F)
5 indicates	§	18	§	47	−40 C (−40 F)
6 indicates	§	18	§	47	− 50 C (−58 F)

Not specified means that no particular elongation is required or that the requirements of the first digit are sufficient. § indicates that there is no E43 equivalent to an E51 electrode that could be designated by the second digit.

RR denotes electrode covering type as below:

A = acid, Ar = acid (rutile), B = basic, C = cellulosic, O = oxidising,
R = rutile (medium covering), RR = heavily covered, S = other types

Optional part

170 denotes nominal electrode efficiency in terms of % metal recovery.

3 denotes welding positions in which electrode can be used, as below:

1 all positions
2 all positions except vertical down
3 flat butt weld and horizontal/vertical fillet weld
4 flat
5 flat, vertical down and horizontal/vertical fillet weld

1 denotes the welding current and voltage conditions recommended as below:

symbol	*DC polarity*	*AC min OCV*
0	+ only	na
1	+ or −	50
2	−	50
3	+	50
4	+ or −	70
5	−	70

Table 4.1 (*cont'd*).

6	+	70
7	+ or –	90
8	–	90
9	+	90

AWS classification A5.1 and ASTM classification A233
This method of classifying mild steel electrodes is based on the use of a four-digit number, preceded by the letter E for electrode.
The first two digits designate the minimum tensile strength of the weld metal (in 1000 psi) in the as-welded condition.
The third digit indicates the position in which the electrode is capable of making satisfactory welds.
The fourth digit indicates the current to be used, and the type of flux coating.
For example, the classification of E7018 electrodes is derived as follows:

E = Metal arc welding electrode.
79 = Weld metal with a UTS of 70 000 psi (483 N/mm^2)
1 = Usable in all welding positions.
8 = Basic type of covering with iron powder, AC or DC.

The detail of the classification is shown below:

FIRST AND SECOND DIGITS

E 60xx As-welded deposit, UTS 60 000 psi (414 N/mm^2) min, for E 6010, 6011, 6012, 6013, 6020 and 6027.
E 70xx As-welded deposit, UTS 70 000 psi (483 N/mm^2) min, for E 7014, 7015, 7016, 7018, 7024 and 7028.

THIRD AND FOURTH DIGITS

The third and fourth digits indicate positional usability and flux coating types:

E xx10 High cellulose coating, bonded with sodium silicate. Deeply penetrating, forceful, spray-type arc. Thin, friable slag. All-positional, DCEP only.
E xx11 Very similar to E xx10, but bonded with potassium silicate to permit use on AC or DCEP.
E xx12 High rutile coating, bonded with sodium silicate. Quiet arc, medium penetration, all positional. AC or DCEN.
E xx13 Coating similar to E xx12, but with addition of easily ionised materials and bonded with potassium silicate to give steady arc on low voltage supply. Slag is fluid and easily removed. All-positional electrode. AC or DCEN.
E xx14 Coating similar to E xx 12 and E xx13 types with addition of medium quantity of iron powder. All-positional AC or DC.
E xx15 Lime-fluoride coating ('basic low-hydrogen') type, bonded with sodium silicate. All-positional. For welding high tensile steels. DCEP only.
E xx16 Similar coating to E xx15, but bonded with potassium silicate. AC or DCEP.
E xx18 Coating similar to E xx15 and E xx16 but with addition of iron powder. All-positional AC or DC.
E xx20 High iron oxide coating bonded with sodium silicate. For welding in flat or HV positions. Good x-ray quality, AC or DC.
E xx24 Heavily coated electrode having flux ingredients similar to E xx 12 and E xx 13 with addition of high percentage of iron powder for fast deposition rates. Flat and horizontal positions only, AC or DC.
E xx27 Very heavily coated electrode having flux ingredients similar to E xx 20 type, with addition of high percentage of iron powder. Flat or horizontal positions. High X- ray quality, AC or DC.
E xx28 Similar to E xx18 but heavier coating, and suitable for use in flat and HV positions only. AC or DC.
E xx30 High iron oxide type coating, but produces less fluid slag than E xx 20. For use in flat position only (primarily narrow groove butt welds). Good radiographic quality, AC or DC.

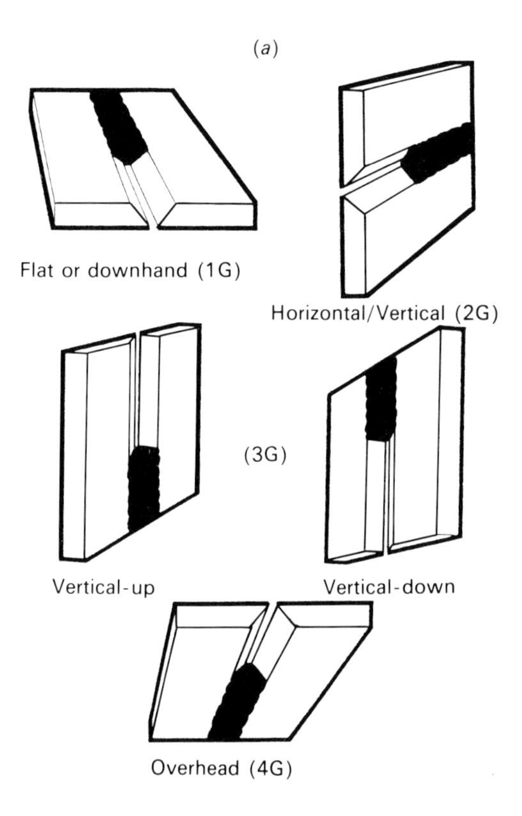

Fig. 4.5 *Standards for the description of welding position (a) British Standard (above) and (b) American Welding Society standard (opposite).*

gives an increase in the speed of welding in terms of both the burn-off rate of the electrode and the deposition rate as Table 4.2 shows.

Electrodes having a recovery rate of at least 130 per cent are sometimes referred to as high-efficiency types. Electrode coverings high in iron powder become partially conducting and are easy to strike.

Cellulosic electrodes

The coverings of these electrodes contain a high proportion of cellulose and leave only a thin slag covering. Decomposition of the cellulose results in hydrogen in the arc atmosphere which raises the arc voltage and therefore there is more 'power' in the arc which is deeply penetrating. The low slag volume makes the electrodes particularly suitable for vertical down and pipe welding using a stove-pipe technique.

Mechanical properties and impact strength are reasonably good and although the weld metal has a relatively high hydrogen content HAZ

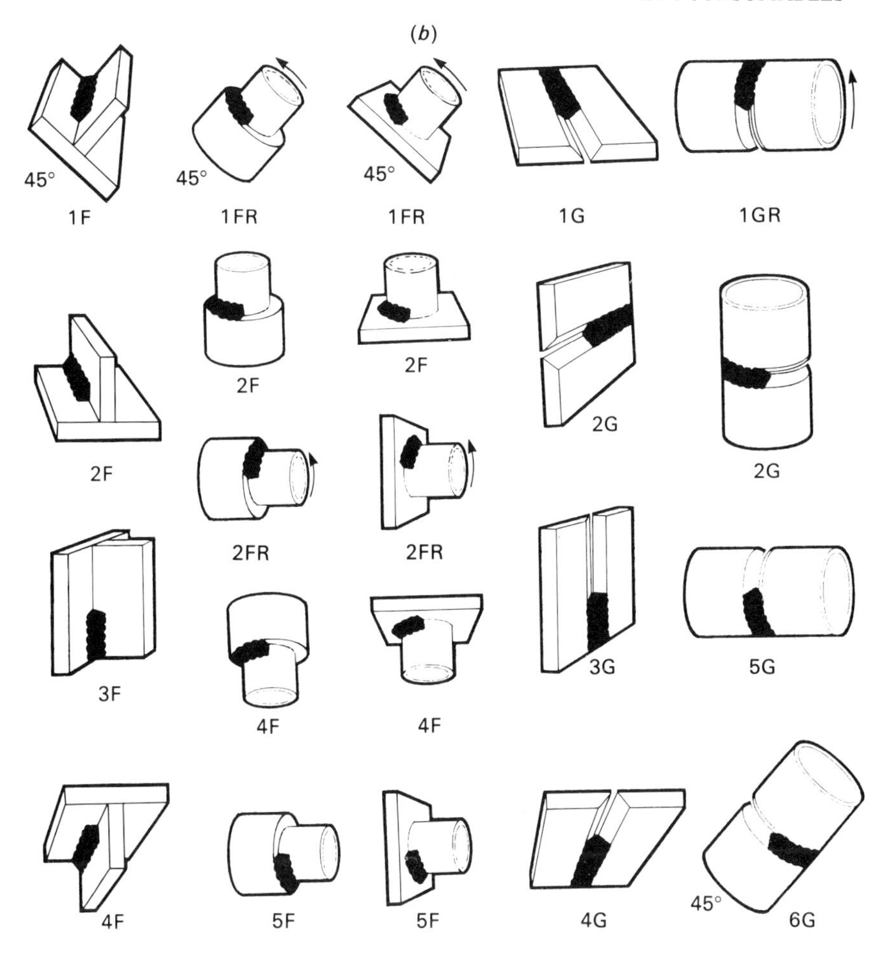

hydrogen cracking is not a problem for the fields in which the electrode is used, i.e. light fabrications and low-strength steels. Avoidance of hydrogen cracking in the higher strength pipeline steels depends on the technique being such as to ensure that second and subsequent runs are in place before the previous run has cooled to cracking temperature. The final or capping phases are put in by weaving to raise the heat input.

Table 4.2 *Comparison of conventional and iron powder electrodes.*

Size	Electrode type	Max current (A)	Time to fuse 100 mm (3.94 in.) electrode (sec)	Time to deposit 100 g (0.22 lb) (sec)
5 mm (0.2 in.)	Rutile	240	25	187
5 mm (0.2 in.)	Rutile iron powder	300	20	77

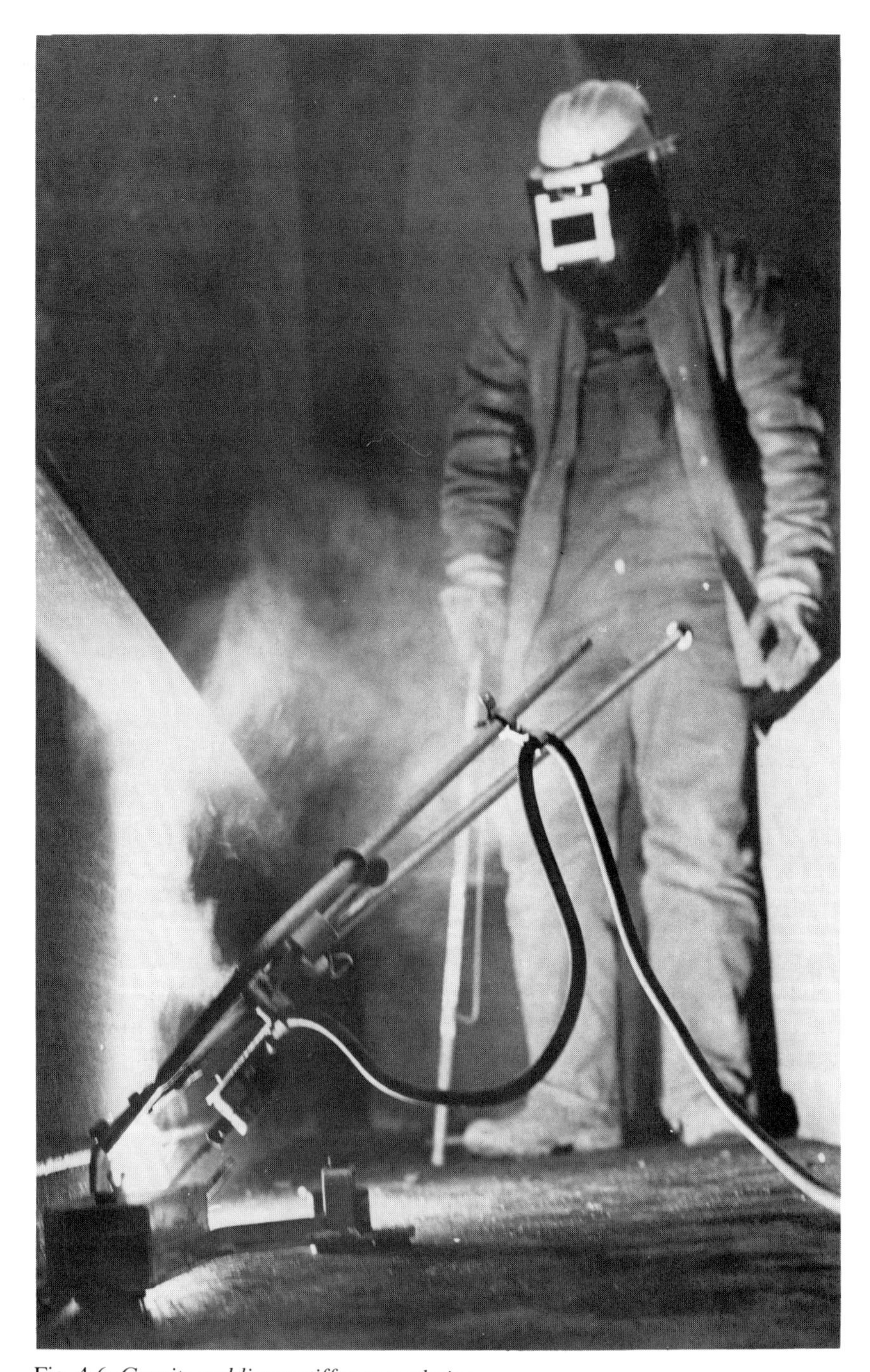

Fig. 4.6 *Gravity welding a stiffener to plating.*

Rutile electrodes

The coverings of these electrodes contain a high proportion of the naturally occurring titanium oxide rutile. This mineral confers on the electrode good recovery because of low spatter loss and easy striking characteristics. Rutile electrodes can work at low open-circuit voltages, giving a smooth finish and are easy to use. This makes them much favoured for tack welding and the welding of sheet steels. They can be used for bridging gaps in preparations and are less sensitive than other electrodes to dirty steel surfaces. Fillet welds made by rutile electronics tend to be of mitre shape, neither concave nor convex. When iron powder is added to the covering improved deposition with metal recovery of up to 175 per cent can be obtained while still retaining the smooth operation and low spatter loss.

Weld metal properties and quality are adequate for general fabrication of mild and low-strength steels but lack the toughness required for higher strength or thick material. The hydrogen content of the weld metal can be high, around 20 ml/100 g, and this is an additional reason why these electrodes are generally unsuitable for higher strength or thick materials.

Gravity welding

The rutile iron powder electrode is particularly suitable for use with the 'gravity welding process' a technique for mechanising MMA welding. (Note that the term 'gravity position welding', meaning welding in the flat position, is sometimes abbreviated to 'gravity welding' with the possibility of confusion. Gravity welding depends upon the electrode burning away to form a deep cup at the end which allows the electrode to be rested on the surface of the workpiece. This technique which can also be practised manually is known as 'touch' welding. It is also essential that the electrode should start easily when the coating is in contact with the work. Two forms of gravity welding equipment exist, in one the electrode holder slides down an inclined rod and in the second the electrode is gripped in a hinge mechanism. Fig 4.6 illustrates the former type. Electrodes for gravity welding are made especially long so there are fewer electrode changes for a given run out length than in manual welding. As the electrode burns at the same rate this is the only speed advantage over manual welding for a single gravity welding unit. The simplicity of the equipment, however, allows several gravity welding units to be operated simultaneously by a single operator and there are then considerable savings. The process is suitable for repetitive work such as the attachment of stiffeners to panels, a common requirement in the shipbuilding and structural engineering industries.

Basic electrodes

These electrodes have in their coverings a high proportion of calcium carbonate and calcium fluoride as the minerals limestone and fluorspar. They have a strong fluxing action together with a carbon dioxide shield and deposit high-quality weld metal with good toughness and low hydrogen, usually 10–15 ml/100 g of weld metal. Special attention to the drying of the electrodes allows this figure to be reduced still further to below 5 ml/100 g for some types. Basic electrodes are therefore specified for the welding of higher strength (above 440 N/mm^2/64 ksi) carbon manganese steels and low-alloy steels also all steels in greater thicknesses when restraint is a problem. They are widely used in the pressure vessel, structural steel and shipbuilding industries. They would be the natural choice for steels with a CE greater than 0.4. (See Chapter 2, page 30, for the definition of carbon equivalent.) Basic coverings are also used for electrodes for stainless steels and hardfacing purposes, and as an alternative to rutile types for specific applications.

The slag from basic electrodes is not so easily detached from the completed bead as that from rutile electrodes. Although the slag melts at a lower temperature it is less oxidising and results in a higher surface tension. For this reason welders find it easier to control the weld pool when welding vertically up and this allows higher welding currents and faster welding than with other electrode types. Also because of surface-tension effects weld beads tend to be convex.

It is particularly important with basic electrodes to maintain a short arc as this ensures that the protection from oxygen and nitrogen by the gas shield is most effective. The generation of the gas shield takes a finite time to complete and hence there is a risk at the start of each new electrode of porosity occurring. Start porosity can be avoided in three ways.

1. Strike the arc 5–6 mm (approx. 0.2 in.) in front of the crater from the previous weld run and move back over the crater before proceeding in the direction of welding. This remelts the start allowing gas to escape.
2. Start about 12 mm (½ in.) back up the previous weld run and move down over the crater. Porosity is contained in the bulbous start which should be ground off.
3. The arc is struck on a piece of scrap plate alongside the joint and once established the arc is transferred to the start of the weld in the joint.

Because the low hydrogen characteristics of the basic electrode are particularly valued these electrodes must be stored in dry conditions and re-dried before use. Moisture can be picked up quickly if the electrodes are exposed to the atmosphere. Many of the new generation of electrodes have coverings especially formulated to avoid hydrogen-bearing minerals and are baked during manufacture to reduce the weld metal hydrogen to less than 5 ml/100 g. The particular advantage of these electrodes being that the very low hydrogen in the deposit makes it possible to relax or eliminate the requirement for preheat.

Over the past 10 years much development has taken place in the field of basic covered electrodes. For example, highly basic coverings providing low weld metal oxygen and good control over the resistance of the weld metal to brittle fracture. Toughness has also been improved by fine adjustments to the balance of alloying elements and by control over the presence of undesirable impurities. Greatly improved welding characteristics now allow single sided welding with good control over penetration, weld bead shape and resistance to the extinction of the arc which occur when the electrode tip comes into contact with the molten pool or parent metal. Improved resistance to moisture pick-up by the coverings is now a characteristic of certain types referred to as MR coverings.

Electrodes for alloy steels

The weld metal from basic covered unalloyed electrodes or the high quality basic electrodes with higher manganese are strong enough for welding a number of low alloy steels. But for welding the more highly alloyed steels, e.g., for creep resisting applications or for use at low temperatures, it is necessary to add ferroalloys to the covering to introduce the required elements to the molten pool.

The flux type may be any of those previously mentioned. Nickel is added to give good low-temperature properties, in particular toughness. It will be seen from Fig. 2.16 that nickel additions of 2–3 per cent have shifted the transition temperature down appreciably. For the most exacting cryogenic applications the nickel content may be increased to as much as 65 per cent. Both nickel and chromium are added for corrosion resistance. Stainless steel electrodes are normally made with a stainless steel core wire although it would be possible to introduce all the alloying elements through the coating. Chromium and molybdenum are added to electrodes for creep resisting applications.

Electrodes for hardfacing range from those depositing a chromium-manganese alloy weld metal with either low carbon or high carbon (the latter can make the deposit difficult to machine) to tubular electrodes depositing either chromium or tungsten carbides in an austenitic matrix. For welding non-ferrous metals or for dissimilar metal and cast-iron welding the electrodes usually have an alloy core wire such as nickel-iron, nickel-copper, pure nickel or copper-tin. Metal arc electrodes for aluminium alloys have alloy core wires and coverings quite different from the electrodes discussed so far. The coverings include alkali halides and leave behind a highly corrosive slag which must be thoroughly removed.

Storage and drying of electrodes

Dampness in some electrode coverings can cause a spattery arc and porosity but its most insidious effect is to introduce hydrogen to the weld and HAZ with the risk of cracking. Users of electrodes should always refer to the manufacturer's literature for the correct procedure to be followed in the storage and drying of electrodes.

Techniques for MMA welding

It is impossible to learn the craft of welding from a textbook alone. A book can, however, indicate general principles and give guidance on what to look for in welding operations.

Welding conditions

The use of welding current values above or below the recommended figures will affect the behaviour of the electrode, and also the appearance of the finished run. If the welding current is too low (see Fig. 4.7), the metal tends to pile up, and the surface of the deposit will be lumpy and irregular. If, on the other hand, excessive current is used (see Fig. 4.8), a flat deposit will be the result, with undue spatter and wastage of the electrode.

The correct arc length is also important. If it is too long the weld metal is deposited, not in a steady stream of small particles, but in large globules which are accompanied by a series of explosions causing spatter and wastage of metal, the rate of deposition is retarded, the weld being irregular and of poor appearance. Welding speed determines the length of deposit from each electrode – this is known as the runout length. Each electrode has a natural run length at which it works best. Table 4.3 illustrates this for fillet welds.

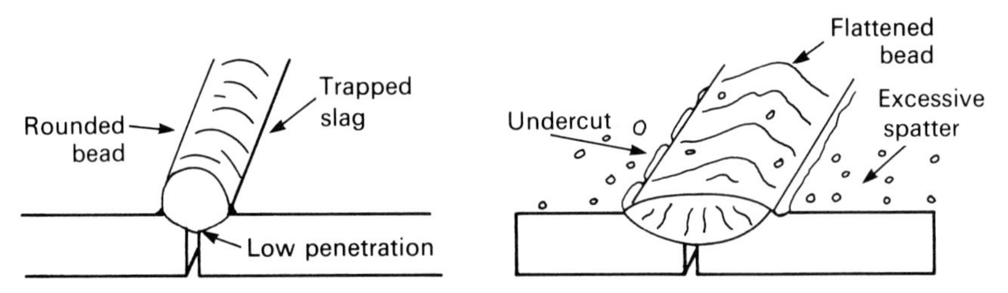

Fig. 4.7 *Weld bead contour when current is too low.*

Fig. 4.8 *Weld bead contour when current is too high.*

Table 4.3 *Typical leg lengths for various electrode sizes.*

Fillet leg length mm (in.)	Electrode size mm (in.)	Approx. runout length mm (in.)
4.0 (0.16)	3.25 (1/8)	300 (12)
5.0 (0.20)	4.0 (5/32)	300 (12)
6.0 (0.24)	5.0 (3/16)	300 (12)

If the runout length is well above normal the electrode may be travelling so fast that there is not time for proper fusion to take place. The weld metal freezes rapidly in the form shown at the left-hand side of Fig. 4.9. When the runout length is short the arc stays too long in one place, causing the metal to run down on to the horizontal plate. Because of the thickness of the run, the quantity of slag will also be excessive and will become uncontrollable (see the right-hand side of Fig. 4.9). In some critical applications the runout length may be specified to produce a particular energy input per unit length of weld to control the cooling cycle and ensure either freedom from hydrogen cracking (see Chapter 2) or optimum mechanical properties.

Large fillet welds are built up in several runs (see Fig. 4.10). Generally

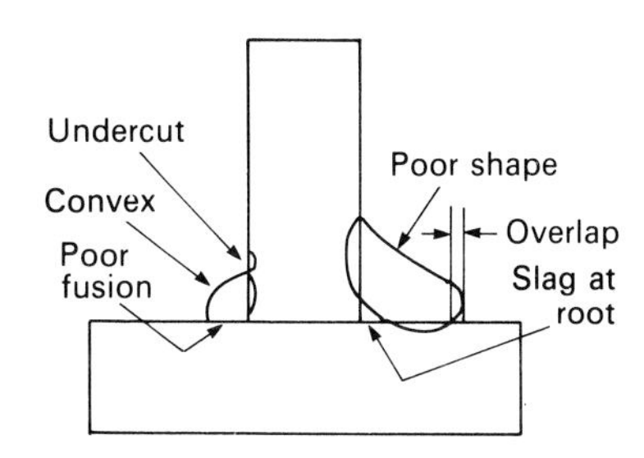

Fig. 4.9 *Effect of runout length on weld profile – (*left*) runout too long, (*right*) runout too short.*

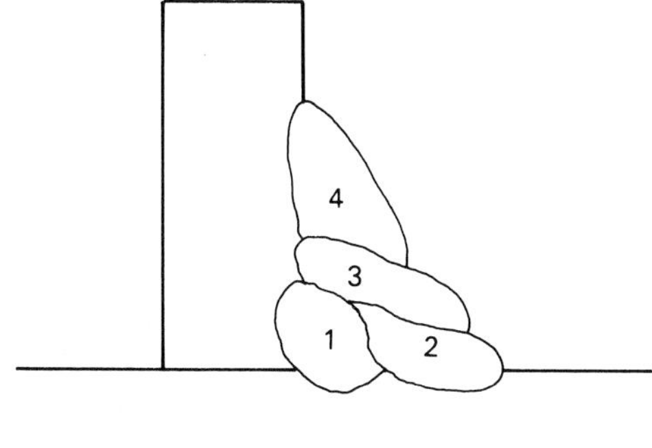

Fig. 4.10 *Arrangement of weld runs in a multirun fillet weld.*

multi-run horizontal–vertical fillets have each run made with the same size of electrode and the same run lengths.

Edge preparation

Material with a thickness of 1.0–3.2 mm (0.04–0.125 in.) can be welded by butting the two parts together. Above this thickness it is impossible for the molten pool to penetrate to the root of the joint. A groove is therefore necessary and this also accommodates the deposited metal which would otherwise form a pronounced weld bead. Typical edge preparations for steel plate are shown in Fig. 4.11.

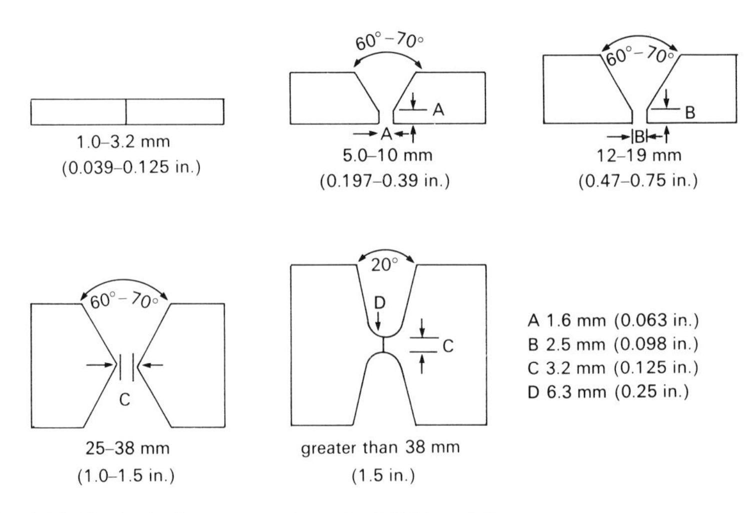

Fig. 4.11 *Typical edge preparations for MMA welding.*

Tacking

The object of tack welds is to hold the assembly to be welded in alignment and to keep any gap at the correct opening. Tack welds are small welds made with a high welding current to obtain proper penetration and have a high cooling rate. Twisted or springy material must be levered into position and additional tack welds made to hold the parts securely. On thick plates where strong tacks are required the tack weld is made in two layers. A weld should never be made over a cracked tack weld as the crack may propagate through the joint. If a tack weld gives way it must, therefore, be chipped out and re-welded. Tacks must be de-slagged before the main weld is made. Tack welding of assemblies for fillet welding is carried out in the same way as for butt welding, but the pitch of the tack welds can be about twice that used for butt joints.

Defects caused by faulty technique

Undercut

This weakens the joint and creates a slag trap (see fig. 4.12).

Cause High current or arc too long. Angle of electrode incorrect – it should not be inclined less than 45 degrees to the vertical face. Joint preparation does not allow the correct electrode angle. Electrode too large for joint. Insufficient time depositing metal at edge of a weave. Weaving is more likely to produce undercut than a straight run, therefore when possible use straight runs.

Incomplete penetration

A gap left by failure of the weld metal to fill the root (see Fig. 4.13).

Cause Current too low or electrode too large for joint. Insufficient gap. Angle of electrode too inclined. Incorrect sequence.

Slag inclusions

These are non-metallic particles trapped in the weld metal (see Fig. 4.14).

Fig. 4.12 *Defects in welds – undercut.*

This weakens the joint and creates a slag trap.

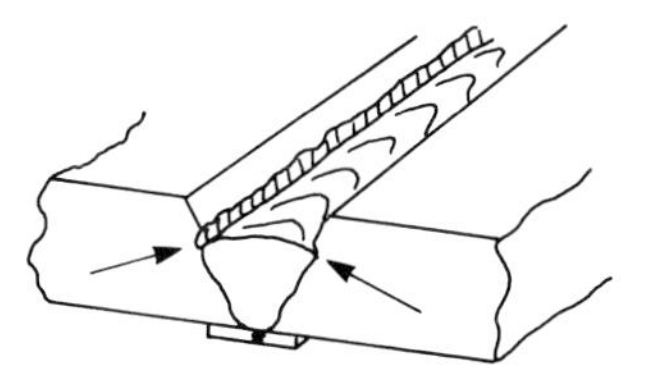

Cause	Remedy
High current.	Reduce amperage.
Arc too long.	Keep shorter arc.
Angle of electrode too inclined to joint face.	Electrode should not be inclined less than 45 deg to vertical face.
Joint preparation does not allow correct electrode angle.	Allow more room in joint for manipulation of electrode.
Electrode too large for joint.	Use smaller electrode size.
Insufficient depositing time at edge of weave.	Pause for a moment at edge of weave to allow build-up. (Weaving is more likely to produce undercut than a straight run. Therefore, where possible, use straight runs.)

A gap left by failure of the weld metal to fill the root.

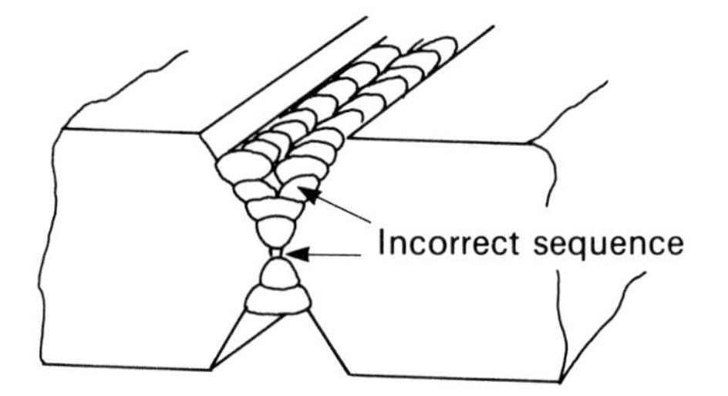

Cause	**Remedy**
Current too low.	Increase current.
Electrode too large for joint.	Use smaller electrode.
Insufficient gap.	Allow wider gap.
Angle of electrode.	If too inclined, does not give penetration. Keep nearer to right angle to weld axis.
Incorrect sequence.	Use correct build-up sequence.

Fig. 4.13 *Defects in welds – incomplete penetration.*

These are non-metallic particles trapped in the weld metal and may seriously reduce the strength of the welded joint.

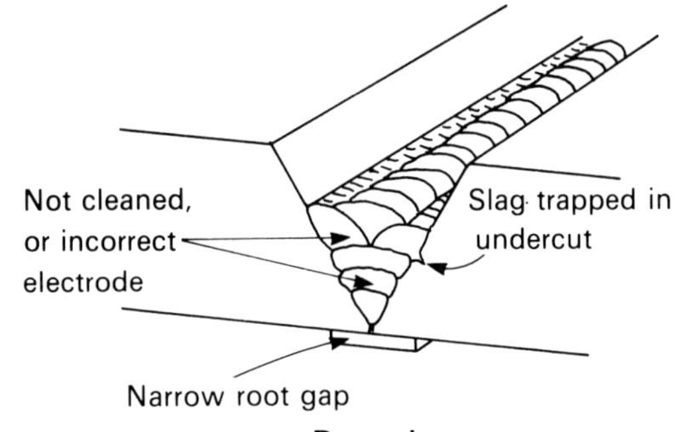

Cause	**Remedy**
May be trapped in undercut from previous run.	If bad undercut present, clean slag out an cover with run from smaller size electrode
Joint preparation too restricted.	Allow for adequate penetration and room cleaning out slag.
Irregular deposits allowing slag to be trapped.	If very bad, chip or grind out irregularities
Lack of penetration with slag trapped beneath weld bead.	Use smaller electrode with sufficient amperage to give adequate penetration. Use suitable tools to remove all slag fro corners, etc.
Rust or mill scale, preventing full fusion.	Clean joint before welding.
Wrong electrode for position in which welding is done.	Use electrodes designed for position in which welding is done, otherwise prope control of slag is difficult.

Fig. 4.14 *Defects in welds – slag inclusions.*

Cause May be trapped in undercut from previous run. Joint preparation too restricted. Irregular deposits allowing slag to be trapped. Lack of penetration with slag trapped beneath weld bead. Use smaller electrode with sufficient current to give adequate penetration. Rust or mill scale, preventing full fusion. Wrong electrode for position in which welding is done.

Lack of fusion

Portions of the weld run do not fuse to the surface of the parent metal or previously deposited weld metal (see Fig. 4.15).

Cause Small electrodes used on thick cold plate. Current too low. Electrode not directed enough towards parent plate. Welding speed too high. Scale or dirt on the joint surface.

NOTE. In overcoming these faults, it is often an advantage, if the job can be positioned to allow welding to be done in the flat position.

Fig. 4.15 *Defects in welds – lack of fusion.*

Portions of the weld run do not fuse to the surface of the metal or edge of the joint.

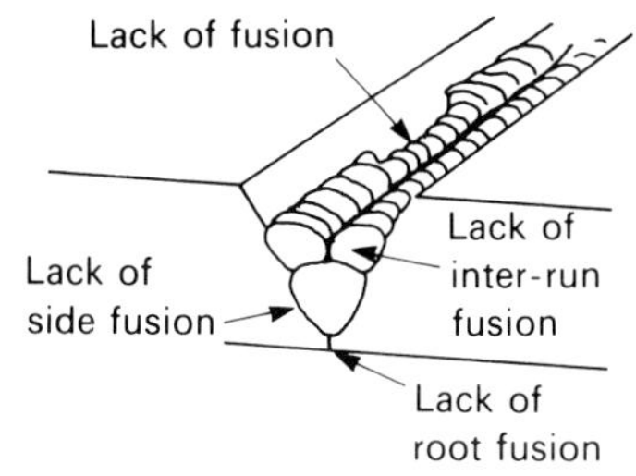

Cause	**Remedy**
Small electrodes used on heavy cold plate.	Use larger electrodes (preheat may be desirable).
Current too low.	Increase current.
Wrong electrode angle.	Adjust angle so that arc is directed more into parent metal.
Speed of travel.	If too high, does not allow time for proper fusion.
Scale or dirt on joint surface.	Clean surface before welding.

Note: In overcoming these faults, it is often an advantage if the job can be positioned to allow welding to be done in the flat position.

Process assessment

The manual metal-arc process (MMA) is the most widely used process for welding ferrous materials and for hard-surfacing operations. Electrodes are available for most types of steels and also copper base alloys. With tubular electrodes material can be deposited with hard particles for wear-resisting applications. The equipment is simple and cheap and can be readily transported and used in inaccessible positions. In the hands of a skilled welder high-quality welds can be produced. The main applications are structural steelwork, process plant, shipbuilding, general engineering, repair and maintenance.

Because the process is intermittent and manually operated it has been replaced in some applications by gas-shielded or flux-cored arc welding which are continuous and offer better productivity. It remains, however, the most important general-purpose welding method.

Chapter 5

The MIG process

Frequently used names for the process – MIG, metal inert gas; MAG, metal active gas; GMA, gas metal-arc; CO_2 welding.

One of two gas-shielded arc processes developed in the 1940s, this process was known originally in the United States as gas metal arc (GMA) and also as MIG (metal inert gas) but later when CO_2 was used for shielding the term MAG (metal active gas) was also used since this gas is not inert and reactions take place between the gas and the liquid pool. Gas metal arc is, however, a good generic title, but in Europe the term MIG is most frequently encountered. The process was used at first chiefly for welding aluminium alloys with helium or argon as the shielding gas. Except for special applications the equipment and shielding gas were at first too costly to enable the process to compete with manual metal arc welding for the welding of steel. This situation was changed some twenty years after the process was introduced by the development of lower cost power sources and shielding by carbon dioxide or carbon dioxide/argon mixtures. The process is now in wide use for welding a variety of materials both ferrous and non-ferrous.

The essential feature of the process is the small diameter electrode wire which is fed continuously into the arc from a coil (see Fig. 5.1 and Fig. 5.2). As a result of the small wire diameter current density is high and the rate at which the wire is consumed in the arc (the burnoff rate) is generally several metres (hundred inches)/min and upwards. The wire is fed at a constant speed and the arc length remains steady because of an electrical feature known as self-adjustment. This will be described in detail later but can be defined briefly as a mechanism by which a change in arc length (giving a change in arc voltage) causes a change in arc current which alters the burnoff rate in a way to oppose the original arc length change. Steadiness in operation is ensured by the burnoff rate being considerably higher than the rate of change of position as a result of involuntary movement by the operator's hand.

The simplicity of the arc length control mechanism and the fact that the small diameter electrode wire can be pushed down a flexible conduit make the process ideally suited for semi-automatic operation when it is known as manual GMA or manual MIG. That is, the welding head or gun on the end of the flexible conduit can be hand held and manipulated but all other operations are automatic. Although the process is ideal for use in this semi-automatic manner it may also be used with great advantage when the gun is manipulated by a machine (mechanised MIG) and consequently it is the welding process most used with arc welding robots, a field of growing importance. The range of materials which can be welded by MIG is almost unlimited being restricted only by the characeristics of metal transfer from wire to workpiece. The equipment for MIG welding comprises the power source which is DC (normally used on the electrode positive polarity), the wire feed unit, the flexible conduit with torch and control devices to start and stop the wire feed, welding current and gas flow. Other controls to improve operation are also incorporated.

A specialised vertical form of MIG welding exists known as electrogas welding. The gun with an automatic weaving unit is mounted on a carriage and moved up the joint and as with electroslag welding, see page 131, water cooled shoes are used to support the rather large molten pools.

Fig. 5.1 *MIG welding process.*

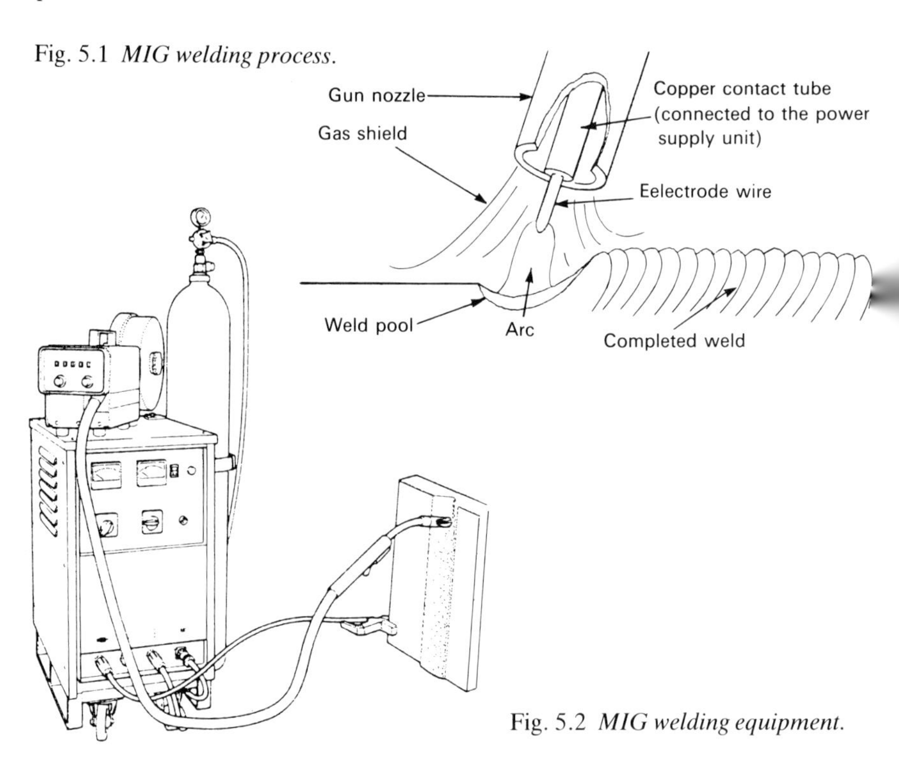

Fig. 5.2 *MIG welding equipment.*

Metal transfer

General

The way in which metal is transferred from the electrode wire to the molten pool for any given material and wire diameter depends upon current, voltage and shielding gas composition. For an aluminium alloy and an inert gas shield with low current, metal transfer takes place as large drops detached by gravity from the wire, a condition unsuitable for welding. As the current is increased the form of the transfer changes abruptly to a stream of fine drops which are propelled across the arc gap by the electro-magnetic forces in the arc. This is called spray transfer and it enables welding to be carried out against gravity. Spray transfer also occurs with steel wires in an inert gas but although the transferred metal can be directed against gravity the molten pool is usually too fluid to be stable. Changing the shielding gas with steel wires to carbon dioxide causes the transfer to become more globular and less well directed; however, the situation can be reversed by using a mixture of inert gas and carbon dioxide.

Decreasing the arc voltage markedly and also reducing current (by reducing the wire feed rate) with steel wires results in a form of transfer known as dip transfer or short-circuit transfer in which metal is fused directly into the pool without passing freely across the arc gap. At slightly higher voltages the transfer is across a gap but is in larger globules without the pronounced directionality of the spray transfer. The globular to spray change is less marked with steel than with aluminium alloys and welds in steel are sometimes made in which this type of transfer predominates. It is also possible to control the type of metal transfer at low to medium currents by using a special power source which delivers pulsed current to the arc. The types of metal transfer and their fields of application are summarised in Table 5.1.

Table 5.1 *Modes of metal transfer for steel wires in an argon 20% CO_2 shielding gas.*

Transfer type	Arc volts	Current (amps)	Applications
Dip	13–23	40–210	Light gauge material, all positions
Spray	24–40	200 and over	High deposition rates on heavier plate and sections, flat and HV only
Globular	20–26	200–280	Higher deposition rates than dip-transfer with lower heat input than spray transfer
Pulsed arc	16–26	60–220 38–50*	For good results on light gauge materials including mild and stainless steels, also aluminium and alloys with argon

*background current

Dip transfer (or short circuiting arc)

In the dip transfer range a low current and voltage setting are used to produce a short circuiting arc. When welding commences and a molten pool is formed, the tip of the electrode dips into the pool and causes a short circuit. This results in a rapid temperature rise in the wire (caused by the short circuit current flowing through to the workpiece) and the end of the electrode is melted off. An arc is immediately formed between the tip of the electrode and the molten pool. This arc maintains the electrical circuit and produces sufficient heat to keep the molten pool fluid. The electrode continues to feed and the tip once again dips into the pool. This sequence of events is repeated at a frequency of up to 200 times per second (see Fig. 5.3).

This method of transfer is suitable for positional welding and has the advantage that the heat input to the workpiece is kept to a minimum. This limits distortion and enables thin sheet material to be welded. Dip transfer occurs with both CO_2 and argon/CO_2 mixtures of shielding gases. Suitable electrode diameters range from 0.6–1.2 mm (0.024–3⁄64 in.) but 0.6 mm and 0.8 mm (0.024 and 1⁄32 in.) are the most used sizes for the dip transfer welding of thin steel. Dip transfer is not generally used with aluminium wires mainly because the wires have too low a resistance to fuse reliably during the short circuit.

Fig. 5.3 *Sequence of events in dip transfer welding.*

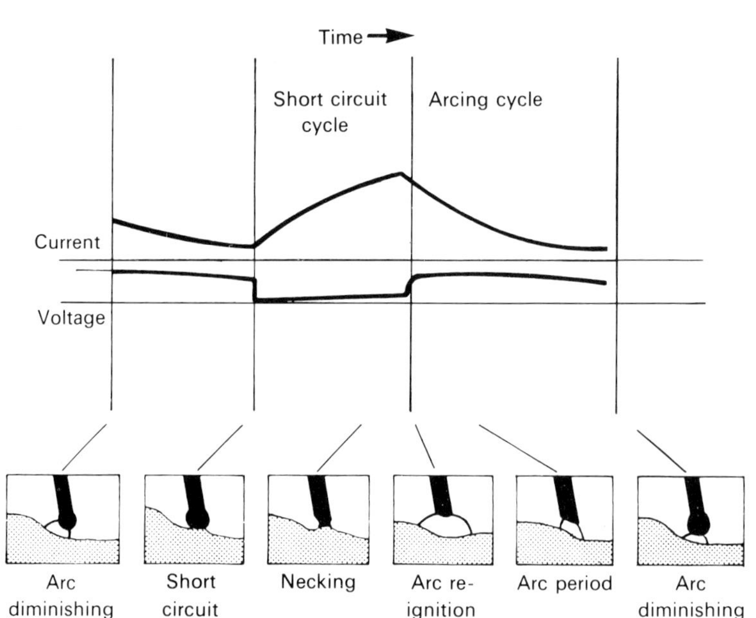

Spray transfer

In the spray transfer mode, higher voltages and currents are used than for dip transfer. The arc tends to envelop the end of the electrode which becomes tapered giving rise to a stream of droplets (see Fig. 5.4). This effect occurs with non-ferrous metals in an inert shield but with ferrous materials it is necessary for some oxygen to be present derived from either additions of oxygen or by dissociation of carbon dioxide. Oxygen up to 5 per cent or carbon dioxide up to 20 per cent are added for this purpose. The size of each droplet is about the same as the wire diameter. Current flows continuously because the high voltage maintains a longer arc and prevents short circuiting taking place. High weld metal deposition rates are possible and weld appearance and reliability are good. Positional welding is not normally possible as the higher heat input makes the molten pool too fluid. Spray transfer is best achieved with argon-based shielding gases and suitable solid wire diameters are from 0.8 mm up to 1.6 mm (1⁄32 in. up to 1⁄16 in.).

Globular transfer

This is an intermediate range between dip transfer and spray transfer, the transfer taking place in the form of irregular shaped globules (see Fig. 5.5). These molten globules of metal fall to the molten pool mainly under the action of gravity rather than arc force as with spray transfer. Unlike argon-based shielding gases, CO_2 will not produce a true spray transfer and globular transfer is the nearest that can be achieved. Although satisfactory for many applications globular transfer produces excessive spatter and an untidy looking weld when compared with welds made using spray transfer and argon-based shielding gases. Globular transfer can be used to advantage if a lower heat input than spray transfer is required. Globular transfer can take place using all sizes of electrode wires.

Fig. 5.4 *Spray transfer welding.*

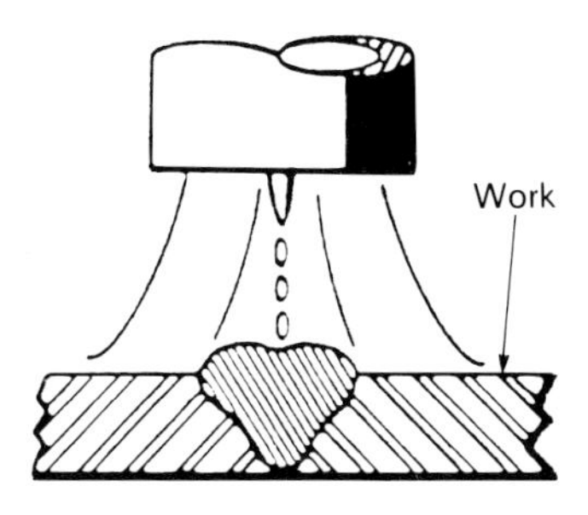

Fig. 5.5 *Globular metal transfer.*

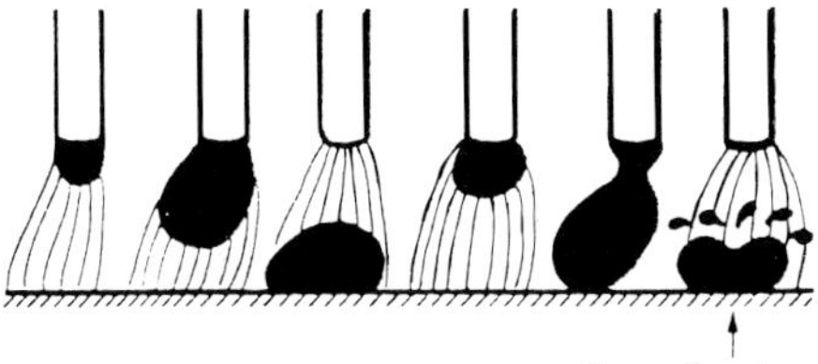

Occasional
short circuiting
brings spatter

Pulsed transfer (pulsed-arc welding)

Pulsed-arc welding is a controlled method of spray transfer welding requiring a more sophisticated power source, whereas the three types of transfer described previously can be obtained with standard power sources and wire feed units. In spray transfer, droplets of metal are projected from the wire tip across the arc gap to the molten pool at a constant current. In dip transfer, metal is transferred to the molten pool somewhat irregularly during the periods of short circuiting. Pulsed-arc welding enables droplets to be projected across the arc gap at a regular frequency, using pulses of current in the spray transfer range supplied from a special power source. Transfer of metal from the wire tip to the molten pool occurs only at the period of pulse, or peak current (see Fig. 5.6). During the intervals between pulses, a low 'background' current maintains the arc to keep the wire tip molten, but no metal is transferred. Pulsed transfer means that the weld metal is projected across the gap at high current, but the mean welding current remains relatively low. The operator can vary the pulse height and the background current to obtain full control of both the heat input and the amount of metal deposited; however, in many modern power sources the pulse procedure is preset by the manufacturer to simplify use. Pulsed-arc transfer can be used on mild and low-alloy steels, stainless steel and is particularly useful with aluminum and its alloys on light to medium plate sections as dip transfer cannot be used on these alloys.

Fig. 5.6 *Sequence of events in pulsed metal transfer.*

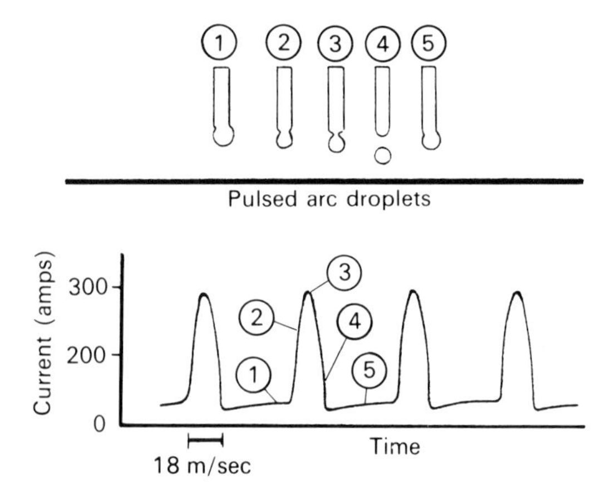

Synergic and controlled transfer welding

The advent of electronic power sources has enabled controlled metal transfer to be achieved with the MIG process. These power sources provide a static volt-amp characteristic that can be tailored to process requirements and a dynamic response significantly more rapid than that from a conventional power source. Feedback from the arc and the wire feeder modifies the power source output to control metal transfer and maintain a stable arc.

High current semiconductors are used to control the welding current employing either series regulator, chopped (switched) secondary or inverter design. With the first two types the mains input is first transformed and rectified, as in a conventional MIG rectifier. With a series regulator the transistors are driven like a valve to control the output current. A highly accurate, ripple-free output with a fast response is obtained, but this form of control is inefficient as considerable power is dissipated in the transistor bank and water cooling is generally required. Transistors driven in the chopped mode offer improved efficiency. They are switched typically at a few kHz to obtain a pulsed output which is then smoothed to give a DC output. Current control is achieved by varying the on to off time of the transistors since the effective current is the average of the pulses. Efficiency is higher, as power is dissipated only during the instant of switching, but performance is generally not as good as with the series regulator.

A recent advance is the inverter power source. This differs from other power sources in that the mains input is first rectified and then switched (by transistors or thyristors) through a welding transformer at high frequency (2–40 kHz), before being rectified again to give a DC output. This technique offers a considerable improvement in efficiency, and performance for the higher frequency inverters is comparable to the series regulator. These electronic power sources can be configured to produce an output specifically suited to a particular process, wire, gas or joint requirement. Feedback from the arc controls metal transfer and maintains arc stability. Metal can be transferred in either the DC or pulsed arc mode.

In DC operation with conventional power sources, at low mean welding currents using dip transfer, the short circuits draw a high peak current which ruptures the contact explosively giving serious spatter. This is followed by a period of low arc current giving a 'cold' weld and the possibility of lack of fusion faults. With controlled metal transfer welding the power source detects that a short circuit is about to occur and injects a high current in a controlled manner. This can control the end of the wire shorting to the liquid pool and will minimise spatter. One type of short-arc technique senses the arc current and wire feed speed. The power source output is adjusted to give a constant arc current per unit volume of wire. As the wire feed speed is varied, the arc current is varied in proportion. This helps to overcome the incidence of lack of fusion faults associated with conventional dip-transfer welding. If should be noted that with these controlled short-arc techniques it is generally the welding current that is controlled (via feedback), rather than the voltage as with a conventional MIG power source.

In pulsed MIG welding (often known as Synergic MIG) a spray-type metal transfer is achieved at low mean welding currents by pulsing the power source output between a low and high level. Metal is transferred only during the peak current pulse. The peak current is chosen to be above the globular to spray transition current level and the duration of the pulse is selected to detach a single molten drop. A low background current is chosen to maintain arc stability and the frequency of pulsing, and hence the mean current, is proportional to the wire feed speed. If, as originally conceived, the power source is operated in the constant current mode, and the pulse parameters are specified in terms of peak current then an additional voltage control of either the wire feeder or power source may be required to maintain arc length. This means that 'one knob' control of the power source can be achieved with the wire feed speed governing the arc parameters.

An alternative system uses arc voltage as the feedback parameter, rather than the wire feed speed. The arc voltage is compared to a reference voltage which corresponds to the required arc length. If the arc length changes and hence with it the arc voltage, the power supply output is modified to restore the required arc length. Again, this control is normally achieved by varying the pulse frequency. The reference voltage should be current-dependent to compensate for the rising characteristic of the arc. With yet other power sources, to maintain a constant arc length, the synergic logic also varies the background current in addition to the pulse frequency. However, if this leads to a change in peak current the drop detachment can be ill-defined. Generally, an additional trim control is required to adjust the arc length and extend the operating range.

Pulsed (Synergic) MIG was originally conceived with the pulse parameters defined in terms of peak and background current, the theory being that current is the main parameter controlling heat input. However, recent power source designs operate with constant voltage in the peak. This gives a measure of self-adjustment during the pulse and offers greater arc stability, particularly with aluminium and cored wires. Metal transfer and arc stability can also be enhanced by generating a more complex waveform than a square pulse. A high-current peak can induce drop detachment and a slope on the trailing edge of the pulse will aid the stable transition between peak and background levels.

Modern electronic power sources therefore enable precise control of metal transfer in MIG welding to be achieved for both DC and pulsed welding. Fusion faults at low mean currents are reduced and spatter levels are considerably lower. Synergic pulsed MIG offers controlled heat input with spray type metal transfer, permitting positional welding with a small weld pool and superior weld finish. (See also the power source section of Chapter 2.)

Equipment for MIG welding

Power sources

MIG welding is carried out on the DC electrode positive polarity (DCEP). Attempts have been made when welding steel to employ DCEN on which polarity the burnoff rate is substantially higher but there are problems stabilising the arc on the electrode. However, DCEN is used to advantage with certain cored wires. Transformer-rectifier power sources are widely used and to enhance the self-adjustment effect these are designed to have a flat characteristic, i.e., they are known as constant potential power sources. Power sources for synergic welding can be constant current types.

The self-adjustment effect, by which involuntary movements of the welding gun by the welder are corrected by changes in the burnoff rate and a steady arc length is obtained, is explained in Fig. 2.2. This compares the current changes with constant current and constant voltage power sources for a given change in arc voltage as a result of altering arc length. Figure 2.2 should be considered along with Fig. 5.7 which shows how the burnoff rate of an electrode varies with a welding current. The constant-current power source, Fig. 2.2(a), is mainly used for manual metal arc welding. Changes in arc length (arc voltage) by the operator have only a minor effect on arc current and this ensures that the weld bead has a uniform shape. The constant potential power source Fig. 2.2(b) used for MIG welding, however, provides a marked change in current with arc length (voltage) which brings about a large adjustment in burnoff rate. The position of the end of the electrode wire with respect to the workpiece depends at any instant on the difference between burnoff rate and wire feed rate. Under equilibrium conditions wire feed rate equals burnoff rate. Shortening the arc causes a decrease in arc voltage, an increase in current and therefore an increase in burnoff rate. The end of the electrode therefore retreats from the workpiece to restore the arc length to its original value as the burnoff rate momentarily exceeds the wire feed rate. Extending the arc length has the reverse effect.

With a constant-current power source the operating current is determined by the intercept of the volt-amp curve on the current axis. This is the short-circuit current which is adjusted by the variable reactor. The operator using such a set for manual metal arc welding takes no account of voltage.

By contrast, when a constant potential power source is used for MIG welding the operating voltage is determined in the first instance by the intercept of the volt-amp curve on the voltage axis. The operator sets the open-circuit voltage slightly above the desired operating voltage. The current drawn by the arc then depends on the wire feed speed setting, the higher the speed the shorter the arc and the higher the current. To establish the operating conditions required (arc voltage and current) the operator has to balance open circuit voltage and wire feed rate.

If it is desired to operate with the dip transfer mode (see page 78) the open circuit voltage must be reduced to give an average arc voltage of around 20 or less (see Table 5.1). When short circuiting takes place during the dip transfer process the current rises rapidly causing the end of the electrode to be fused off into the pool. This fusing must not be too violent or metal will be expelled from the pool as spatter and it is usual

Fig. 5.7 *Welding current–burnoff curve for a steel wire in an argon-carbon dioxide atmosphere.*

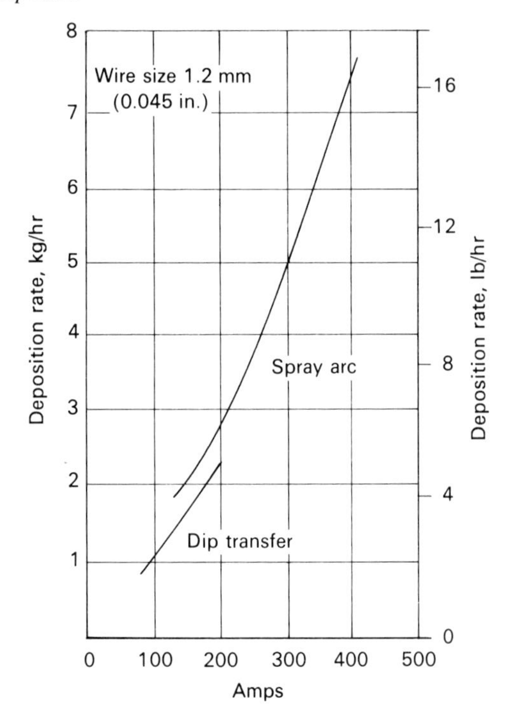

therefore to include in the circuit a variable inductance (or reactor) as was mentioned in Chapter 2, p. 20. This inductance controls the rate of current rise which not only influences the occurrence of spatter but also alters within limits the ratio of arc time to short-circuit time. High inductance gives fewer short circuits because of the slower current rise. Because more heat is put into the weld pool during the arcing part of the cycle high inductance with its greater heat input is used on thicker material where good fusion is required. Conversely low inductance is used when welding thinner material.

The selection of process parameters described above is a procedure somewhat complicated by the fact that the conditions for minimum spatter may not give the desired penetration shape. The procedure must be thoroughly mastered by the operator before good welds can be made. It is often referred to as 'tuning a power source'.

Electronic power sources

Although the traditional power sources have been used with success for many years the development of power sources with refined electronic control systems has greatly simplified the operation and extended the scope of MIG arc welding. In addition, because they frequently incorporate mains voltage stabilisation these power sources overcome the previously intractable problem of transferring the welding conditions established at one site to another situation.

The main advantage of the electronic power sources, however, is that they can be made simple to set up and there is now available more than one system for 'one-knob' control. Electronic power sources may also be preprogrammed so that all the necessary process and welding parameters for a range of jobs can be stored in memory and called up as required. Pulsed arc welding first applied with traditional power sources was too inflexible and difficult to set up. Electronic power sources overcame these limitations and with the development of synergic pulsing have enlarged the scope for MIG. In the synergic system pulse length and current amplitude are selected to enable pulses to be injected to transfer one drop per pulse over the whole operating range.

Aspects of using electronic power sources were discussed in the earlier section on controlled metal transfer, page 80.

Wire feeders

The device for feeding the wire electrode in MIG welding is critical to the operation of the process. It must be capable of pushing the wire down a flexible conduit up to 5 m (15 ft) long, smoothly and with a rapid start. A jerky feed causes arc length and current fluctuations which affect the weld bead shape and quality. Hard wires are easier to feed than small diameter soft wires such as aluminium and types of wire feeders have been developed to cope with small diameter and soft wires which have an auxiliary pulling device incorporated in the welding gun. With small diameter wires there is also the option of the reel-on-gun system, a gun with its own feed rolls and a small lightweight reel of wire carried on the gun itself. Small reels of wire are costly and the guns tend to be awkward to use in some situations so the reel-on-gun has been superseded by other feeding systems. Push–pull systems can allow a welding gun to operate up to 15 m (50 ft) or more from the main feed unit and reel of wire (see Fig. 5.8).

The shorter the feed conduit the easier it is to feed the wire and the less trouble is likely in service; however, the choice of feed conduit length will be determined by the job being done and by how close to the

Fig. 5.8 *A push–pull wire feed system used for welding aluminium alloy road tankers.*

Fig. 5.9 *A self-contained composite power source and feed unit.*

Fig. 5.10 *Heavy duty MIG unit.*

Fig. 5.11 *Wire feed roller system.*

Fig. 5.12 *Tandem wire feed drive.*

operator the feed unit and reel can be placed. For light duty work where welding is to be done within a limited area it is more convenient to have the feed unit incorporated within the same case as the power source (see Fig. 5.9). This is called a composite or compact unit. For heavier duty work over a limited area the feed unit may be mounted on a swivel on top of the power source (see Fig. 5.10). If work must be done over a wide area it is more convenient to have the feed unit completely separate from the power source. When working in inaccessible places a specially compact light-weight feed unit in a 'suitcase' format may be used. Suitcase units are especially useful on complex fabrications where feeding up to 70 m (200 ft) from the power source is required.

The simplest feeding system uses a pair of rollers one of which is usually grooved having an opposing roller which is flat and spring loaded to apply pressure (see Fig. 5.11). To avoid the necessity for changing rollers to suit different sizes of wire the main roller may have more than one groove. Whether one or both rollers are driven is a matter of choice but a dual drive is possibly more positive. When driving a wire for some distance the pressure required on the rollers to maintain a steady feed may be such that the wire is distorted or flattened. This is more likely to happen with soft or cored wires. To overcome this problem the contact area between wire and rollers can be increased either by using a large-diameter roller and several smaller pressure rollers on a portion of its periphery (capstan feed) or by using a second set of rollers in tandem (four-wheel drive). (See Fig. 5.12.) All feed systems have a quick-release mechanism to allow wires to be removed and replaced. On leaving the rollers the wire must immediately enter a guide before passing down the conduit. The feed conduit may be spiral metal for hard wires but for the softer non-ferrous wires a PTFE or plastics low-friction liner is used. The speed at which the wire is fed depends upon its composition and diameter and the welding current being used. Most MIG feeders can handle wires from 0.6 mm to 1.6 mm (0.024–1/16 in.) with a maximum feed speed of about 18 m/min (700 ipm). Further information can be found in the burnoff curves given later.

Controls included within the wire feed unit are a speed-control system usually electronic, for the wire feed motor, solenoid-operated gas valves to turn the flow of gas on and off, a braking system on the drum of wire to stop over-running and various timing devices. One of these timing devices allows current to flow for a short period at the end of the weld to prevent stubbing of the electrode.

Other timing devices are available to allow a complete sequence of starting up and switching off to be carried out within a few seconds. This type of control is used to allow arc-spot welds to be made in overlapping sheets. A special nozzle with a castellated rim is pressed against the upper sheet and the process is fired for a predetermined time by pressing the trigger on the gun. This allows a molten pool to penetrate the upper sheet and fuse into the lower member. More elaborate preprogramming equipment is available which allow combinations of welding variables to be selected readily. Timing devices may be incorporated in the feed unit or an add-on unit or placed within the power source casing.

Push–Pull Systems

Because the flexible conduit is such an important factor in determining the working area special mention must be made of the systems available to allow its extension. There are three such feeding systems as follows:

1. Intermediate.
2. Electric push–pull.
3. Pneumatic push–pull.

In the intermediate system an auxiliary wire feed unit is placed between a gun with a standard length conduit and a wire-feed unit with its conduit but without its gun. The auxiliary unit pulls the wire from the wire feed unit through one conduit and then pushes it down the second conduit to the gun. With the electric push–pull system the normal feed unit pushing the wire is supplemented with a small feeding unit in the gun itself, the handle of the gun containing the motor. Wire-feed rate may be adjusted from a control on the side of the gun. The pneumatic push–pull works in a similar manner except that the feed rolls in the gun are driven by an air turbine in the gun handle.

Welding guns

The device held by the operator is variously called a torch or gun, the latter being the British Standard preferred term. *The International Welding Thesaurus* uses the term 'gun' for a device through which filler wire is fed; hence MIG guns but TIG and plasma torches, see Glossary. The gun has the function of directing the electrode wire and conveying to it the welding current, as well as shrouding both wire and molten pool with the shielding gas. The first two functions are performed by a copper or Cu-Cr-Zr alloy contact tube several centimetres (2 or more in.) long. The working end of this contact tube is within a concentric shielding nozzle some 2 cm (¾ in.) diameter. This performs the third function of directing the gas. The internal design of the gun must be such that the gas emerges from the nozzle in a streamlined flow with a minimum of turbulence. The gas-shielding nozzle becomes hot in use especially when the duty cycle is high and it may be necessary to cool it, or the fitting into which it is fastened, with internal water passages. At high currents appreciable heat is also generated in the contact tube because of the voltage drop across the contact.

As it is the end of the contact tip which suffers most wear it is usual to have replaceable tips. The harder copper alloys are usually employed and the bore allows for a clearance of 0.2–0.3 mm (0.008–0.012 in.) with the wire. Because of slight changes in the straightness of the wire inside the contact tube the point of current pick-up is constantly changing. With high-resistance electrode wires this means that there is a varying amount of resistance (I^2R) heating of the wire. Various methods are employed to force the point of contact as near the end as possible, e.g., by a local reduction in diameter or a stepped and inclined end or a finger against which the wire bears.

There are two main types of MIG welding gun:

1. Those in which the wire and gas pass through the handgrip on the gun. These are generally the lighter duty, air cooled, lower current types and the exit portion has a slight curve from which they derive the name 'gooseneck' or 'swan-neck' (see Fig. 5.13). Heavier duty gas-cooled guns have been designed for currents up to 500 A and these are often preferred to the water-cooled variety because of their reduced weight and improved manoeuvrability.
2. Heavy duty water-cooled guns which have handgrips at an angle to the barrel. In guns of the push–pull type incorporating feed

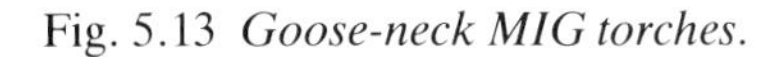
Fig. 5.13 *Goose-neck MIG torches.*

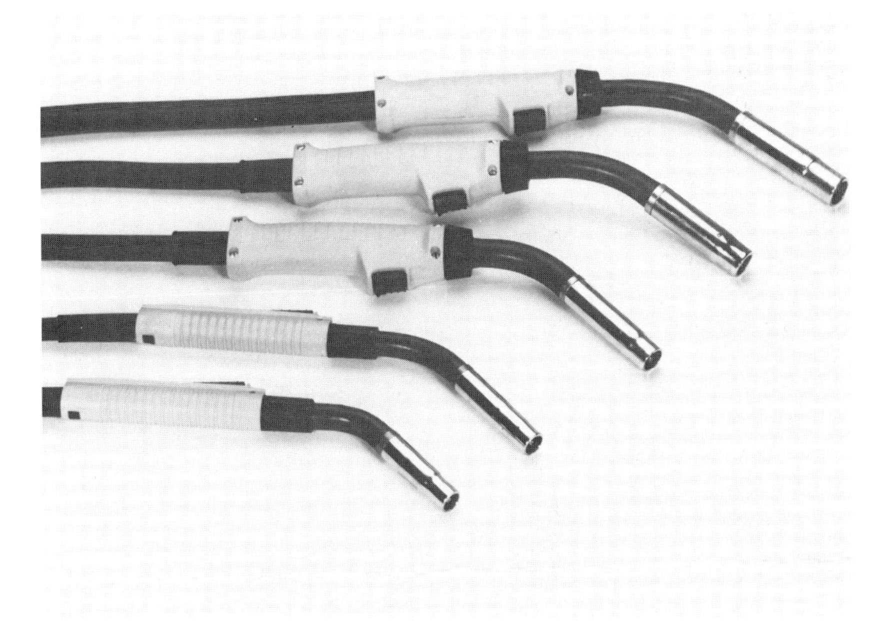

rolls the handgrip is at 90° to the barrel which is usually straight, although a curved end may sometimes be fitted.

The second type of guns sometimes have a curved metal shield between grip and barrel to protect the operator's hand from radiant heat. A switch operated by pressure is provided on the grip of all welding guns to enable the welder to start and stop welding.

Gas supply

Gas for welding is usually supplied direct to each welding station from cylinders. In high-duty welding shops gas is often supplied through a manifold system which allows the cylinders to be stored outside. The change-over to a new cylinder when the one in use is exhausted can be entirely automatic. Argon, helium and the gas mixtures are supplied in cylinders at a pressure of about 200 bar (2900 psig) which requires a regulator to reduce the pressure to a level suitable for welding.

Regulators

A pressure regulator is a device which keeps the delivery pressure relatively constant although the flow rate may be changed and the gas cylinder pressure becomes lower as the supply of gas is exhausted. A single-stage regulator comprises a chamber closed by an adjustable spring-loaded diaphragm. Attached to the diaphragm is a valve which opens or closes an orifice leading to the high-pressure cylinder side of the regulator. When gas is withdrawn by the welding appliance the pressure under the diaphragm is reduced and the valve opens to allow gas through. The output pressure tends to change (increase or decrease according to valve design) as the cylinder pressure drops with usage. A two-stage regulator which is two single-stage regulators in series overcomes this difficulty and allows almost the whole contents of a cylinder to be discharged without a noticeable change in outlet pressure. Pressure gauges are fitted between cylinder and regulator (indicating the cylinder pressure and hence its level of contents) and at the output side of the second diaphragm (indicating the pressure of the supply of the appliance).

For gas-shielded welding it is necessary to know the flow rate of the gas through the shielding nozzle. Flow meters are usually of the type in which gas flow lifts a ball or float in a vertical transparent tube. The graduations on a flow meter apply only to the gas for which they were calibrated but flow meters with multiple scales graduated for several gases are available. When carbon dioxide is used as the shielding gas and is supplied from cylinders it is necessary to fit a heating coil between regulator and cylinder. This is because carbon dioxide in cylinders is liquid and the change of state from liquid to gas causes a drop in temperature which can freeze the regulator.

Consumables for MIG welding

Gases

Argon or argon/helium mixtures are used for non-ferrous metals and nickel alloys. Argon with additions of 5–20 per cent carbon dioxide is used for welding carbon or low-alloy steels and argon with 1–5 per cent oxygen is used for stainless steels where carbon pick-up from dissociated carbon dioxide must be avoided. Carbon dioxide alone is used for dip transfer welding of carbon steels and at higher currents with globular transfer where the higher spatter levels associated with this gas can be tolerated. The gas flow rate for shielding depends on the gas nozzle diameter but is increased as the welding current is raised.

The reasons for the choice of a particular gas are wide-ranging. They may be metallurgical as with the use of inert gases for non-ferrous metals and why argon/oxygen is preferred for welding stainless steel. Generally, however, the choice of gas is determined by the process characteristics, either the need to control spatter or the shape of the penetration of the molten pool. With steels, argon/oxygen mixtures tend at high currents to give a deep, narrow, fingered penetration which can be undesirable as it can be associated with root porosity and in butt joints with the penetration missing the joint at the root. Carbon dioxide gives a slightly higher arc voltage and a hotter arc producing a more spatter than argon-carbon dioxide mixtures (see Fig. 5.14). An argon/5 per cent CO_2 mixture is generally recommended for light gauge work and an argon/20 per cent CO_2 mixture or carbon dioxide for heavier work. Carbon dioxide is possibly less susceptible to lack of fusion defects because of the hotter arc but both gases produce high-quality ferritic weld metal. The oxygen or CO_2 content of the gas may influence

Fig. 5.14 *The effect of shielding gas on metal transfer and penetration.*

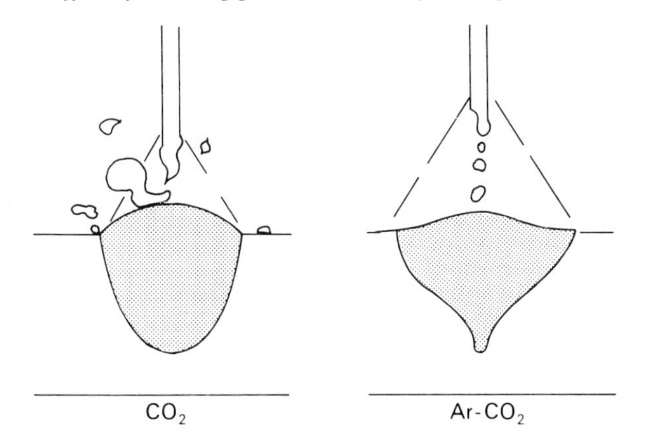

the choice of filler wire as more deoxidants are used up the higher is the available oxygen.

Argon/helium mixtures are used where it is desired to achieve a more bowl-shaped penetration and the use of carbon dioxide is unacceptable. This may occur with the welding of non-ferrous metals where the mixture may be used instead of argon or with stainless steel in place of argon/oxygen mixtures. The process needs protecting from draughts which may disturb the gas shield causing a fluctuating arc voltage, porosity or oxide inclusions.

Wires

Electrode wires for MIG welding are a quality product. They must have a smooth finish to feed easily and be free from metallurgical flaws and both surface and drawn-in contaminants which might give rise to hydrogen, porosity or inclusions in the weld. The wire diameter must be closely controlled as this can affect burnoff rate. Wires are usually supplied on reels and wound to ensure that the wire does not snag when being withdrawn. Both random wound and layer wound products are used, the layer wound being the more costly. The physical properties of the wire are critical to its ease in handling and feeding. If the wire is too soft it may be difficult to push down the conduit causing buckling and a jerky feed. When too hard the wire becomes springy which also gives a jerky feed. If it has a twist the wire will gyrate on leaving the contact tube causing the arc to wander. Each material and size of wire requires different treatment.

Specifications for wires are to be found in BS 2901: 'Filler rods and wires for gas-shielded arc welding', Part 1: Ferritic Steels; Part 2: Austenitic Steels; Part 3: Copper and copper alloys; Part 4: Aluminium and aluminium alloys and Magnesium alloys; Part 5; Nickel and nickel Alloys. American Welding society A5.18 Carbon Steel; A5.28 Low alloy; A5.9 Stainless steel; A5.7 Copper and copper alloys; A5.10 Aluminium and aluminium alloys; A5.14 Nickel and nickel alloys; A5.16 Titanium and titanium alloys. Also in ISO 864 carbon steels.

Ferrous metals

Steel wires for general use are given a coating of copper during the final stages of drawing to improve the current pick-up properties and to protect the wire surface from corrosion. This copper finishes up in the weld metal where it is usually regarded as harmless. There may be circumstances where copper must be avoided, however, and non-copper coated wires are available. For high-current, high-duty cycle welding non-copper-coated wires are available. For high-current, high-duty cycle welding non-copper-coated wires are sometimes preferred because it is believed that copper can flake off the wire and accumulate in the flexible conduit causing feed problems. Non-coppered wires are specified by about 20 per cent of users. The most widely used steel wire is a carbon steel deoxidised with manganese and silicon. Another frequently used wire is deoxidised with aluminium, titanium and zirconium (the so-called triple deoxidised wire). This is recommended where steel may be rusty and for pipe welding and the root runs of thick joints. Where more strength is required for welding higher tensile steels, e.g., as in pipeline welding a wire containing 0.5 per cent molybdenum is often used. Wires for creep-resisting applications containing chromium and molybdenum are also available.

Deposition rates vs. current curves for carbon-manganese wires in argon/CO_2 and CO_2 are similar. For metals of low electrical resistance there is a straight-line relationship between burnoff rate (and therefore deposition rate) and current. Steel and higher resistance metals show an increasing burnoff rate with current because of resistance heating (also called the I^2R effect), which is accentuated with the smaller diameters of the wire. Variations in the length of wire protruding from the contact tube, known as electrode extension or stickout, therefore affect deposition rate as do quite small changes in the composition of electrodes and wire surfaces. Wire diameter also affects the lower limit at which dip-transfer welding can be carried out and the current at which it is necessary to change to spray transfer, as well as the maximum current. All are raised by increasing the wire diameter (see Table. 5.2). Another effect of wire diameter at the same current is to change the shape of the penetration bead which becomes narrower and more penetrating with smaller diameter wires. The most used sizes of wire are the 1.0 and 1.2 mm (0.039 and 3⁄64 in.) diameter wires.

The dip-transfer process can be used in all welding positions on all types of joint; however, it is not normally allowed for the positional welding of plates thicker than 5 mm (0.2 in.) in applications where it is essential to guarantee freedom from lack of fusion defects. It is a tolerant process and can accommodate the presence of unintended gaps

Table 5.2 *Current limits for different wire diameters.*

Wire diameter	mm (in.)	0.8 (0.032)	0.9 (0.036)	1.2 (0.048)	1.6 (0.064)
Current for dip					
Min.		40	50	60	100
Max.		170	190	200	210
Current for spray					
Min.		—	—	125	220
Max.		225	260	400	450

in the preparation of joints. When gaps are being bridged it may be helpful to increase the stickout by withdrawing the gun slightly. The increased resistance-heating decreases the average current and gives a cooler deposit. Vertical-down welding should be limited to material no more than 6 mm (0.236 in.) thick to avoid cold laps forming. Vertical-up welding used for fillet welds in metal over 6 mm (0.236 in.) thickness tends to give a convex bead and weaving is necessary to limit the extent of this. Overhead welding requires a circular weave to flatten the bead. Flat and horizontal welding require a leftward (forehand) techique with weaving as appropriate. An important use of the dip transfer technique is for the all-position welding of pipelines either manually or with orbital welding machines. It is also widely used for sheet metal welding as in the construction and repair of automobiles and is convenient for assemblies such as lattice girders, where a large number of short welds must be made and the absence of forced electrode changes is an advantage. The absence of slag (except for the small islands of glassy slag which appear as a result of deoxidation reactions) reduces finishing time and gives clean neat results. The ability to weld fast or to deposit small fillets reduces distortion.

Welding in the spray regime which is limited to the flat and horizontal positions is used for applications in the 3–25 mm (⅛–1 in.) thickness range. High efficiency iron powder electrodes have competitive deposition rates but the MIG process may still be preferred because it is continuous, deeply penetrating, allows narrower edge preparations and produces high quality weld metal low in hydrogen. Table 5.3 indicates the welding conditions which might be used for MIG welding a number of different joints in steel.

Stainless steels

A pure argon gas shield does not give ideal metal transfer with stainless steels and may also result in incomplete penetration. An argon/1–2 per cent oxygen mixture is often used to overcome these difficulties.

Early experience with MIG welding of thick stainless steels using argon/1–2 per cent oxygen mixtures was often unsatisfactory because porosity appeared in the root of the deeply fingered penetration bead. The solution to this problem was found in the use of argon/helium or argon/2 per cent CO_2 gas mixtures. At a level of CO_2 as low as 2 per cent carbon pick-up is not a problem when welding low carbon grades and as the carbon dioxide mixture may be the cheaper of the two gases this is preferred. When buttering on a ferritic steel base or when welding the higher carbon stainless steel an argon/5 per cent CO_2 mixture can be used. Matching electrode wires for the common grades of stainless steel are available. Some wires are made with a slightly higher silicon content

Table 5.3 *Examples of welding conditions for Argon-CO_2 MIG welding of steel.*

Technique	Thickness of work mm (in)	Wire dia. mm ($\frac{in.}{1,000}$)	Amps	Volts	Wire feed m/min (ipm)
Dip, V	1 mm B (0.04)	0.8 (32)	90	15	5.25 (206)
Dip, V	3 mm F (0.12)	0.8 (32)	155	17	8.25 (325)
Dip, V	3 mm B (0.12)	1.0 (40)	160	18	7.00 (275)
Dip, V	6 mm F (0.24)	1.0 (40)	160	18	7.00 (275)
Dip, V	12 mm F (0.47)	1.0 (40)	175	19	9.25 (365)
Dip, V	20 mm MRF (0.79)	1.0 (40)	175	19	9.25 (365)
Dip, V	12 mm F (0.47)	1.2 (48)	175	18	3.8 (150)
Dip, Fl and V	20 mm MRB & F (0.79)	1.2 (48)	175	18	3.8 (150)
Spray, Fl	3 mm B (0.12)	0.8 (32)	165	25	12.0 (473)
Spray, Fl	3 mm F (0.12)	0.8 (32)	175	25	12.9 (508)
Spray, Fl	3 mm B (0.12)	1.0 (40)	230	26	13.45 (530)
Spray, Fl	6 mm F (0.24)	1.0 (40)	210	27	12.9 (508)
Spray, Fl	6 mm F, B (0.24)	1.2 (48)	320	32	10.5 (413)
Spray, Fl	12 mm MRF (0.47)	1.2 (48)	320	32	10.5 (413)

B = butt weld
F = fillet weld
MRF = multi run fillet
MRB = multi run butt
Fl = flat position
V = vertical position

than the parent metal to improve the welding properties in respect of wetting and surface finish.

Aluminium alloys

The introduction of the MIG process was largely responsible for the development of aluminium alloys for structural applications requiring welding. Prior to its introduction only gas welding, manual metal arc and gas tungsten arc were available. Gas welding and manual metal arc required active fluxes based on chlorides and fluorides to remove the tenacious refractory surface oxide film and the processes were difficult to use with metal arc giving poor weld appearance.

Gas welding was suitable for thin gauge butt welding of the pure metal and the low alloys but the removal of the corrosive flux residues presented a problem. The high thermal conductivity of aluminium was a bar to gas welding being used on thick material and fillet welding was impossible because of difficulties in securing penetration and in removing trapped flux. Fillet welds could be made in thick material by manual metal arc but preheating was often required. The process was difficult and unpleasant to operate becaues of spatter and poor bead shape and flux residue removal was also difficult. There was also an intractable problem with weld metal porosity because the hygroscopic fluxes were impossible to free from moisture. Gas tungsten arc welding overcame the flux residue problem and gave excellent weld soundness but was unsatisfactory for fillet welding and the heat was not sufficiently concentrated to allow the welding of thick material.

All these difficulties were overcome by MIG which proved to be particularly suitable for the fillet welds which form a significant part of any structural welding application. Important early applications were the superstructures of ships, vessels for the chemical industry, and the bodies of large transport vehicles.

A lower limit is placed on the current which can be used for any wire diameter by the occurrence of globular metal transfer. When the current is increased transfer changes suddenly from globular to spray and may be increased until the upper limit is reached when the deeply penetrating arc causes turbulence in the pool and the entrainment of air into the arc column. This gives rise to an oxidised and deeply puckered weld bead. The lower current limit for spray transfer depends on wire diameter and to make the process applicable to thinner material smaller diameter wires may be used. Wires with diameters from 0.8 mm to 2.4 mm (1⁄32–3⁄32 in.) are currently available. Being relatively soft these wires are difficult to push down a flexible conduit and require careful control in manufacture as well as in use. The conduit and contact tube should be well maintained and the conduit should not be subjected to severe bending. Push–pull feed systems are normally recommended. Pulsed welding is particularly suited to aluminium alloys and allows thicker wires to be used at lower currents with excellent shaped weld beads. The deposition rate is independent of wire diameter, in contrast to steel. Aluminium wires are given a special chemical or shaving treatment during the final stages of manufacture to remove the hydrated oxide film acquired in processing and replace it with a thin uniform oxide film less susceptible to moisture pick-up during storage. Ths is necessary to prevent hydrogen entering the weld pool where it results in porosity. Finer electrode wires having a greater surface area for a given mass of metal give welds more susceptible to this type of porosity than thicker wires.

Electrode positive (DCEP) is always used for aluminium and its alloys with either an argon or argon/helium shield. Wires are available in a range of compositions to suit applications, the commonly available compositions being aluminium, both with and without titanium as a grain refiner, used to weld pure aluminium for chemical plant and electrical conductors; aluminium-5 per cent silicon for welding the Al-Mn and low strength heat treatable alloys of the Al-Mg-Si type; aluminium-3 per cent magnesium for the lean aluminium-magnesium alloys and aluminium-5 per cent magnesium for the higher alloyed aluminium-magnesium alloys of the shipbuilding type. The aluminium-5 per cent silicon alloy is not used as an electrode wire on the aluminium-

Table 5.4 *Examples of conditions for welding aluminium alloys*

Joint thickness mm (in)	Electrode diameter mm ($\frac{in.}{1,000}$)	Wire feed rate m/min (ipm)	Welding current amp
1.5 (0.060)	0.8 (32)	8.5 (335)	80
3.0 (0.118)	1.0 (40)	8.9 (350)	130
6.0 (0.236)	1.2 (48)	9.3 (366)	200
12.0 (0.472)	1.6 (63)	7.0 (275)	250
Arc voltage 20–22			

magnesium alloys because although the weld metal is resistant to weld cracking the presence of much Mg_2Si causes brittleness and poor corrosion resistance. Where a weldment must be subject to weathering or may be anodised the aluminium-5 per cent silicon electrode gives a poor colour match, the weld becoming grey. The aluminium-5 per cent magnesium electrode gives a better colour match but is not so resistant to weld cracking. The high-strength heat treatable alloys are not arc welded. Table 5.4 indicates the welding conditions which might be used for butt welding four different thicknesses of aluminium alloy.

Copper and copper alloys

Copper and its alloys are widely used for their corrosion resistance and thermal and electrical conductivity. The large range of alloys provides numerous combinations of these properties with different strengths and there is a range of filler metals and wire electrodes for the many possible applications. The introduction of the gas-shielded processes overcame many of the problems experienced with welding copper and copper alloys by gas and metal-arc. The high heat input of the MIG arc was particularly valuable in allowing thicker sections to be welded without preheat. Argon and argon/helium mixtures are used for shielding. Helium with its ability to raise arc voltage and heat input helps to reduce the need for preheating with the high conductivity parent metals such as tough-pitch, phosphorus-deoxidised and oxygen-free coppers. The latter two coppers are those generally welded because the residual oxygen in tough-pitch copper can cause porosity. When welding unalloyed copper it is usually necessary to employ an electrode wire containing small amounts of a deoxidant such as manganese, silicon, titanium or boron.

Three copper alloys are commonly welded, copper-silicon (silicon bronze), copper-aluminium (aluminium bronze) and copper-nickel (cupro-nickel). The alloying elements assist deoxidation of the molten pool but the filler metal or electrode wire usually contains additional deoxidants to ensure complete freedom from porosity. Copper-zinc

Table 5.5 *Examples of conditions for MIG welding copper and copper alloys.*

Alloy	Thickness mm (in)	Preheat C (F)	Welding current	Arc voltage	Wire feed rate m/min (ipm)
Copper	6 mm (0.236)	None	240–320	25–28	5.5–6.5 (216–256)
	12 mm (0.472)	up to 500 (930)	320–380	26–30	5.5–6.5 (216–256)
	18 mm (0.71)	up to 500 (930)	340–400	28–32	5.5–6.5 (216–256)
Cu-Si	6 mm (0.236)	None	250–320	22–26	6.0–8.5 (236–335)
	12 mm (0.472)	None	300–330	24–28	6.0–8.5 (236–335)
	18 mm (0.71)	None	330–400	26–28	6.0–8.5 (236–335)
Cu-Al	6 mm (0.236)	150 (300) max.	280–320	26–28	4.5–5.5 (177–216)
	12 mm (0.472)	150 (300) max.	320–350	26–28	5.8–6.2 (228–244)
	18 mm (0.71)	150 (300) max.	320–350	26–28	5.8–6.2 (228–244)
Cu-Ni	6 mm (0.23)	150 (300) max.	270–330	22–28	4.5–5.5 (177–216)
	12 mm (0.47)	150 (300) max.	350–400	22–28	5.5–6.0 (216–236)
	18 mm (0.7)	150 (300) max.	350–400	24–28	5.5–6.0 (216–236)

All welds made using 1.6 mm (1/16 in.) diameter wire and argon shielding.

alloys are welded infrequently because of the heavy zinc fume which is evolved. This can be reduced by using an electrode wire such as aluminium bronze, the aluminium present forming a film over the molten pool. Filler metals for all the alloys mentioned are available. With single phase (6–8 per cent Al) aluminium bronzes a tendency to root embrittlement and cracking is avoided by using a non-matching two phase alloy with higher aluminium and additions of iron, nickel or manganese. This alloy, however, can suffer from a defect known as de-aluminification by corrosive action and may require a surface layer of parent metal composition. Copper-nickel alloys are sensitive to oxygen and hydrogen contamination during welding and gas shielding must be carefully maintained with additional argon shielding used on the under-side of the weld. Electrode wires specially developed for copper-nickel alloys containing titanium are necessary. The high conductivity coppers are used in electrical applications and together with aluminium bronze and copper-nickel alloys are used in pipework and heat exchangers and marine fittings. Table 5.5 indicates the welding conditions which might be used for butt welding three different thicknesses of copper and copper alloys.

Process assessment

MIG welding with solid wires is an extremely versatile process. With argon mixtures or carbon dioxide shielding it allows the welding of thin to thick steel in all positions. It is particularly suitable when used with inert gases for welding non-ferrous metals, notably aluminium alloys which can be welded in all positions. The process provides easy arc striking and weld control with continuous welding and no electrode changes. The welds are clean, requiring no deslagging. There is less distortion than with manual metal-arc and narrower edge preparations are possible. It is a low hydrogen process giving high quality weld metal.

The equipment is readily mechanised or installed on robot welding plant but it is relatively expensive although ranging in price. Regular maintenance and a supply of spare parts is necessary. Knowledge other than that relating to manipulative skill is required of the operator in setting up and use. There is a requirement to protect the working area from draughts and the equipment is not so portable or easy to use in restricted spaces as manual metal-arc. The process is, however, capable of higher productivity than manual metal-arc.

Chapter 6

Cored-wire welding

Frequently used names for the process – FCAW, flux cored arc welding; tubular wire welding

The cored-wire process

With this process, often known by the abbreviation FCAW, a cored wire comprising a metal tube filled with flux and metal powder is used with the same basic equipment as for MIG welding with solid metal wires (see Fig. 6.1). Many cored wires are used with a gas shield of either carbon dioxide or argon based mixtures but some types contain within the core a flux and other materials which alone give adequate protection from the atmosphere and these, the self-shielded wires, can be used without additional gas shielding. The carbon-manganese types of self-shielding wire are often employed in heavy structural steel applications on site, others deposit stainless steel or are used for

Fig. 6.1 *Principle of cored wire welding.*

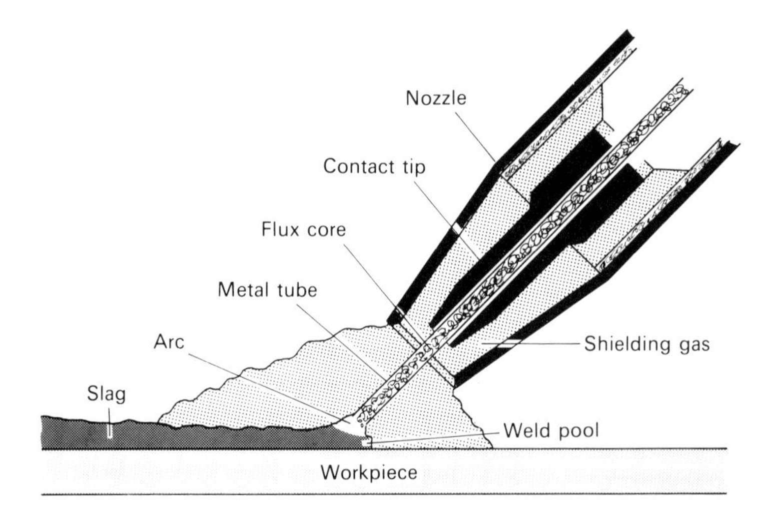

hardfacing, when the alloying elements can be placed more conveniently in the core rather than alloyed into a solid wire which would be difficult to draw down to wire. The tubular metal sheath is usually mild steel but can be made from other metals depending on the intended use of the wire.

Cored wires are generally used on thicker material, e.g., about 4 mm (0.16 in.) upwards, than the solid wires used with the MIG process. The fact that molten pool protection and the control of weld metal composition can come from both gas and wire can give added assurance of weld quality. Cored wires are more satisfactory than solid wires when light mill scale or primers have to be tolerated. Cored wires can be used at higher welding currents than solid wires and when advantage is taken of this it is possible to obtain considerably improved deposition rates and productivity. The fact that the welding current is conducted almost entirely through the outer sheath makes a cored wire more susceptible to I^2R heating than a solid wire of comparable size. The current density in the sheath is greater than with solid wires, providing a faster burnoff rate, and the arc is also more dispersed which gives good side wall fusion. The presence of a slag on the surface of the molten pool from the flux core assists with positional welding.

The powder filling of a cored wire fulfils the same functions as the covering of a metal arc covered electrode, i.e., protection from atmosphere, deoxidation, slag forming, arc stabilisation and alloying addition. Certain cored wires contain mainly metal powders with a small amount of deoxidants and arc stabilisers. These metal-cored wires do not have a slag to protect the weld metal but retain most of the advantages of flux-cored wires, such as high deposition rate, though to a lesser extent than the flux cored wires.

When fluxes are used for metal-arc covered electrodes binders must be present to give strength to the covering. These binders are a source of hydrogen and to obtain a low hydrogen deposit electrodes must be baked at elevated temperatures. Binders are unnecessary with cored wire which can therefore have a lower hydrogen potential, approaching that of a solid wire. Generally, the need to rebake low-hydrogen MMA consumables to guarantee low-hydrogen weld metal is not a requirement for cored wires in which the flux is protected by the outer sheath of metal, and providing wires are stored in dry conditions and rusting is avoided this low-hydrogen potential is maintained. Modern cored wires are superior to the earlier product in respect of hydrogen potential.

When first introduced cored wires were only available in diameters considerably greater than the solid wires used for MIG welding. With the development of the butt-closed wires, rather than overlapping strip, it has become possible to draw wires down to 1.2 mm (3⁄64 in.) diameter or less. Popular sizes are now 1.2, 1.4, 1.6 and 2.4 mm (3⁄64, 0.055, 1⁄16 and 3⁄32 in.) diameter. Because of the method of manufacture cored wires are considerably more expensive than solid wires. In countries other than the USA the utilisation of cored wires has been slow in developing but more recently the process has been adopted increasingly in Japan. Consumable cost which initially may appear to be a controlling factor is not of overriding importance, however, as other factors are also important, such as the higher usable current, easier positional welding, improved penetration, reduced sensitivity to wind and the more ready availability of a range of compositions than with solid wires. For nearly three decades the usage of cored wires in the USA has been 15 per cent or more of all weld metal deposited, but since 1980 the proportion of cored wire usage has been growing steadily in the rest of the world and this trend will continue.

Equipment

Power sources

Any power source designed for use with solid wire MIG welding will be suitable for cored wires, with the important proviso that it must have a high enough maximum current output to cope with the higher welding currents generally used with cored wire. With a 2.4 mm (3⁄32 in.) diameter wire a current capacity of 500/600 A is required, the higher capacity being needed for mechanised operation at a 100 per cent duty cycle. A wire of 1.6 mm (1⁄16 in.) diameter will normally require at least 400 A capacity for manual welding and 500 A if the operation is mechanised.

Cored wires are almost invariably used on direct current; DCEP (electrode positive) for rutile cored and hardfacing wires, but DCEN can be used for basic and metal cored wires. Although metal transfer is slightly different from that for solid wires and also varies from one type of cored wire to another the full range of techniques from dip and spray to pulse is available. The range of power sources in use on cored wires is therefore as broad as that for solid wire. Rectifier sets require an adjustable choke to permit dip transfer welding and use is also made of forms of pulsed welding and pre-programmed electronic control. The latter is particularly relevant when cored wires are used to avoid lack of sidewall fusion in positional welding, a common problem when solid wires are used in MIG welding. Welding sets are available in which all these control developments are available in a single unit. Such versatility is not always a luxury as cored wires exhibit a wide range of behaviour. Many further process benefits will be achieved when the metal transfer characteristics of highly basic or self-shielded wires are improved through the use of rapid-response electronically controlled power sources. Wires of the self-shielding or hardfacing type are often used on site when normal mains power is not readily available and in this situation engine-driven sets can be used in conjunction with specialised wire feed units.

Wire feeding systems

The reliable use of cored wires requires a special attention to the feeding roll system. Increasing the roll pressure to ensure a good grip may result in the wire becoming deformed with the consequence of feeding difficulties in the flexible conduit or contact tube. While the smaller diameter 1.2 or 1.4 mm (3/64 or 0.055 in.) wires may be fed using the grooved drive roll and flat pressure roll commonly used for solid wires it is normally preferred, especially for the larger sizes, to employ knurled drive rolls geared together. Other more elaborate drive systems have been developed, the most widely used of which is a tandem roll system. Although tandem rolls may be driven by a single motor a twin motor drive can also be employed. Electronic speed regulation, a remote control option and the facility to interface with robot and mechanised systems, all of which are found increasingly on MIG equipment, are also desirable. See MIG welding, page 85, for wire-feed system.

Welding guns

In choosing any welding gun for either solid or cored wires account must be taken of the operating current, which will depend on the size of the wire, and the duty cycle. Cored wires are run at higher welding currents than the same size of solid wire and water-cooled guns are often required for 1.6 mm (1/16 in.) wires and sizes larger than this. Water-cooled guns are generally more costly than their gas-cooled counterparts, however, and much design work has been done on the provision of gas-cooled high duty guns capable of a duty cycle of 60 per cent at 500 A, even when using argon-rich shielding gas.

The liner inside the feeding conduit is usually of spirally wound steel but for stainless steel wires a Nylon or Teflon liner may be used. Conduits are normally 3–4 m (10–13 ft) long but where a longer length is necessary to provide the required mobility two or more lengths can be connected using intermediate in-line feed units. There is a risk when using excessive feed roll pressure of finely divided metal becoming detached in the liner causing feeding difficulties and liners should be checked regularly for the presence of swarf.

When self-shielding wires are used the gun can be simplified, a gas nozzle not being required. This is an advantage when working in positions of difficult access and it also permits the use of a long electrode extension which increases the control which can be exercised by the welder over I^2R heating and penetration. This is especially useful for hardfacing for which self-shielding wires are common.

Consumables

Gases

For gas-shielded cored wire welding of ferritic steels with normal carbon levels the choice of gas is usually between carbon dioxide and argon 15/20 per cent CO_2 mixtures. Carbon dioxide gives good coverage and protection and requires an arc voltage about 2 V higher than an argon-based gas, and metal transfer tends to be globular. Ar/CO_2 mixtures with which spray transfer can be obtained readily result in less fume, less spatter, smoother weld deposits and better 'tie-in', i.e., smoother transition from weld to plate. Argon-based mixtures must always be used when welding with metal cored wires. With the same wire, the loss of alloying elements will be less and the weld metal strength therefore higher for an argon-based gas than for CO_2. For welding stainless steel and high alloy materials where carbon contents must be kept to a minimum Ar/CO_2 mixtures can increase the carbon level and Ar/O_2 shielding gas mixtures may be necessary. Gas flow at the nozzle should be in the range 15–20 litres per minute (L/min) (32–42 cfh) for flux-cored wires and 20 L/min (42 cfh) for metal cored types.

Wires

The cored wire process uses a metal tube made from strip which is formed to enclose a dry powder mix. The material of the strip is usually mild steel but can be stainless steel for stainless wires or even copper if the wire is to deposit aluminium bronze. This illustrates the versatility and potential for cored wire welding.

Four main configurations are in use for the tube, seamless, butt closed, overlap, and formed, the latter having various degrees of overlap or more complicated folding (see Fig. 6.2). The form of the tube influences the amount of powder which can be contained per unit length or the 'fill ratio'. Seamless wires generally contain the least powder, butt-closed

Fig. 6.2 *Four types of cored wire:* (a) *seamless,* (b) *butt closed,* (c) *overlap and* (d) *formed.*

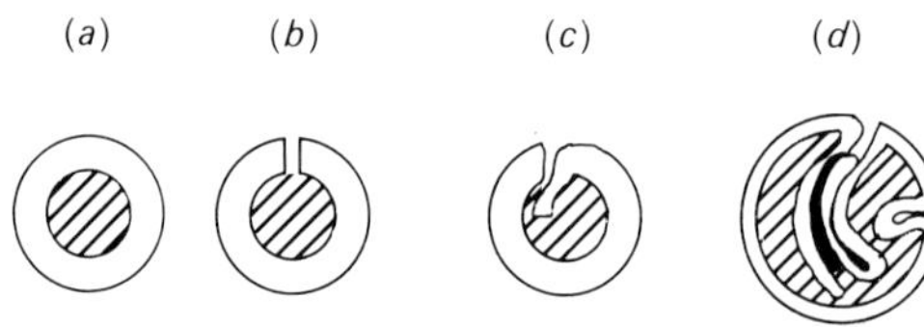

wires are intermediate and overlapped wires the greatest powder fill. The high fill ratio of the overlapped type can offer advantages for surfacing wires and formed complex sections are often used for producing large diameter wires of say 4 mm (5⁄32 in.) and above. The most commonly employed cored wires are of the butt-closed type having a reasonably high fill ratio as well as being suitable for drawing to small diameters. Special non-metallic coatings have been developed for cored wires to make them easier to feed and to give resistance to rusting in storage.

Cored wires may be gas-shielded or self-shielded. Gas shielded wires are either flux-cored, the filling being rutile (titanium oxide) or basic (containing calcium carbonate and calcium fluoride) with possibly some iron powder in either case, or they may be metal-cored with a filling of metal powder along with deoxidants. Self-shielded wires generally have a rutile base with extra deoxidants, denitriders and gas-forming compounds. The main options and their operating conditions are set out in Table 6.1.

Many of the considerations discussed in Chapter 4 under the heading of covered electrodes (MMA) also apply to flux cored wires. Rutile-cored wires are easy to use, give a smooth finish and flat standing fillets with satisfactory properties for general use. Rutile-cored wires are usually less easy to use for positional welding although special flux formulations giving rapid freezing to allow positional welding are available. Root runs with rutile-cored wires require backing.

For superior low temperature properties and when welding thick material a basic wire is required. Enhanced strength and low-temperature performance are obtained with further additions of nickel and molybdenum to a basic flux core. Basic cored wires work well in dip transfer and can therefore be used at low currents allowing positional welding and the welding of root runs.

When welding creep resisting steels a basic core with additions of

Table 6.1 *Summary of cored-wire welding consumables.*

Wire type	Self-shielding	Gas shielded		
		Flux core		Metal core
		Rutile	Basic	
Shielding gas	None	CO_2* Ar/CO_2	Ar/CO_2† only	
Electrode polarity	DCEP/AC	DCEP	DCEN/DCEP	DCEN/DCEP

*May be limited to dip transfer condition up to 200 A.
†Normal European practice.

chromium and molybdenum is employed to give matching composition weld metal (except for vanadium-containing steels where the wires do not contain vanadium, as low vanadium weld metal is desirable to avoid stress relief cracking).

Wires for welding stainless steels have a rutile/basic core containing the necessary alloying elements often within a tube made from low carbon mild steel. Some manufacturers prefer to use a stainless steel as with smaller diameter wires which have a low fill ratio is it difficult with mild steel to introduce enough alloying elements. Certain stainless steel cored wires can be used with or without gas shielding equally satisfactorily but metal deposited without a gas shield will generally have a higher nitrogen content which decreases the ferrite content. Nitrogen content is also increased by using long arcs. These factors may need to be taken note of in applications where ferrite control is required to avoid solidification cracking.

A wide range of wires is available for hardfacing to give deposits with a hardness of 280 up to 750 HV for components subject to abrasion, impact, heat and corrosion. The alloying elements are usually contained wholly within the core and the wires may be gas-shielded rutile type, metal cored or self-shielding. Self-shielding wires with a rutile flux are widely used and advantage is taken of the self-shielding property to use a longer electrode extension, up to 35 mm (1.38 in.), than would normally be possible. This gives good access and according to how conditions are set either a higher deposition rate or lower heat input. Wires with a diameter of 1.6 mm (1⁄16 in.) are used for positional work and 2.4 mm (3⁄32 in.) diameter wires for work in the gravity position.

Standards

The most widely used standards relating to cored wires at present are those of the American Welding Society 'Specification for carbon steel electrodes for flux cored welding, AWS A5.20', which uses a combination of letters and numbers as follows:

1. The letter E indicating an electrode.
2. A number indicating the strength in units of 10,000 p.s.i.
3. A number indicating the welding position for which the electrode was designed, 0 for F and HV and 1 for all positions.
4. The letter T indicating a cored wire.
5. A suffix denoting usability and performance characteristics as follows:

   (*a*) – electrodes used with CO_2 or argon-CO_2 shielding for single and multipass welding. Most have a rutile-based slag, give spray transfer, low spatter loss, flat to slightly convex bead and a moderate slag volume.

(*b*) – electrodes that are essentially similar to type 1 but with higher manganese or silicon for single-pass welding and horizontal fillets over scaled or rimmed steel.
(*c*) – electrodes are self-shielding, DCEP, welding in flat, horizontal or downhill positions, single pass welds, with spray transfer. High welding speeds are possible, but these wires should not be used for multipass welds or material thicker than 4.8 mm (3⁄16 in.).
(*d*) – self-shielding electrodes that operate with DCEP and globular transfer. They give high deposition rates, low penetration, and weld metal that is very resistant to cracking. They can be used in joints with poor fit-up, and for single or multipass welding in the flat and horizontal positions.
(*e*) – electrodes are designed for CO_2 or argon-CO_2 shielding for single pass or multipass welds in the flat or horizontal positions, and for horizontal fillets, globular transfer and slightly convex beads. The thin lime-fluoride slag may not entirely cover the weld. The weld metal has higher impact and crack resistance than rutile types.
(*f*) – self-shielding, DCEP, spray-type transfer. The slag system is designed to give good low temperature properties, deep penetration beyond the weld root, excellent slag removal in deep grooves, and the electrodes are suitable for single pass and multipass welding in the flat and horizontal positions.
(*g*) – self-shielding and DCEN. Smaller electrode sizes may be used in all positions and the larger sizes give high deposition rates in single pass or multipass welds, with good resistance to cracking.
(*h*) – self-shielding, DCEN, all-position welding and good low-temperature impact properties in single pass and multipass welds.
(*i*) – classification is not used.
(*j*) – self-shielding, DCEN, high travel speed capability, single pass welds in the flat, horizontal and downhill positions.
(*k*) – self-shielding, DCEN, smooth spray-type arc. These electrodes are designed for all-position welding and high travel speeds in single-pass and multipass welds.

The specification for low alloy cored wires (AWS A5,29) follows the same general arrangement as for the carbon steel electrodes except that the usability and performance classification becomes part of the main description. The suffix now indicates the composition of the deposited weld metal, e.g., A1 is a C-Mo steel; B2 a Cr-Mo steel; Ni1 a C-Mn with 1 per cent Ni, K a high-strength steel.

For stainless steel wires the AWS A5.22 specification uses the letter E for electrode followed by a three digit number indicating composition, followed by the letter T indicating a cored wire and finally a suffix denoting the shielding method, 1 means CO_2, 2 means Ar/O_2 and 3 means self-shielding.

A detailed review of international specifications for welding consumables is published by The Welding Institute.

Techniques for cored-wire welding

Setting welding conditions

To get the best performance and productivity from a cored wire it is important to understand how it differs from solid wire and covered electrodes. Figure 6.3 compares the deposition rate of several cored and solid wires with rutile covered electrodes. It shows that at any one welding current the deposition rate is higher for smaller diameter wires. This effect which is more pronounced with cored wires is a result of resistance heating of the electrode which contributes to the overall melting rate and means that 1.2 mm (3⁄64 in.) diameter solid wires have a higher deposition rate at any current than 1.6 mm (1⁄16 in.) diameter solid wires which have almost twice the cross-sectional area. Cored wires only conduct current through a part of the cross section and a 2.4 mm (3⁄32 in.) flux-cored wire will only have about two-thirds of the current carrying section of a 2.4 mm (3⁄32 in.) solid wire. For any welding current therefore cored wires give a considerably higher deposition rate

Fig. 6.3 *Deposition rates for various steel wires and processes.*

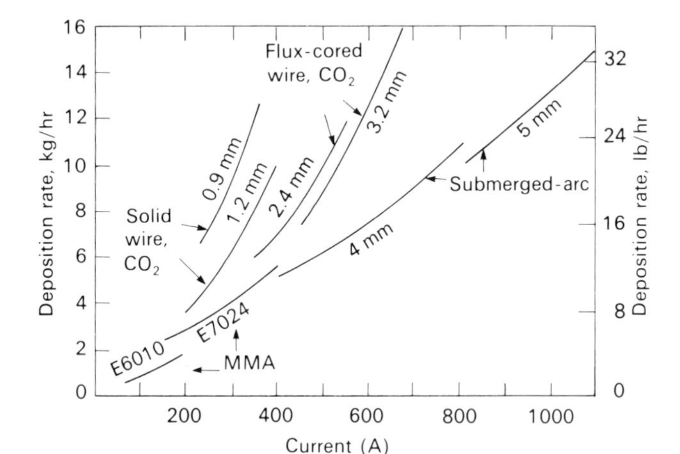

than the same size of solid wire. The deposition efficiency of flux-cored wires is, however, lower at 79–86 per cent compared with about 95 per cent for solid wires, the loss occurring by spatter and slag formation. The deposition efficiency of metal-cored wires, however, can be as high as 95 per cent.

Because of the presence of a slag which holds the molten pool in position and because the arc root with flux-cored wires is more distributed cored wires can be operated at higher welding currents than solid wires. The usable currents with metal-cored wires having no slag are much closer to those for solid wire. The higher welding current gives deeper penetration and if the design of the structure allows account to be taken of actual throat depth rather than theoretical throat depth this is a further bonus for covered wires. The productivity advantage of cored wires is most apparent when welding in position but productivity advantages can also be demonstrated in many applications when welding in the flat or gravity positions. The effect of inappropriate welding conditions in causing weld faults is summarised in Table 6.2 at the end of this section.

Welding current

With flux-cored wires the current should be within the top half of the range specified by the manfacturer except when using dip transfer with smaller wires when the current would normally be no higher than 220 amps. As with solid wires penetration depends primarily on welding current. Metal-cored wires can be used at single current setting for almost all plate thicknesses when making flat or HV welds. The weld cross-section is determined by the welding speed. Table 6.3 indicates the range of welding current used for different wires and wire diameters.

Table 6.2 *Rectifying weld faults.*

Fault	Possible cause
Undercut	Speed too slow or gun angle too low or voltage too high.
Lack of penetration	Current too low or electrode extension too great or joint preparation too narrow or root face and gap too small.
Lack of fusion	Voltage too low or direction of travel wrong (leftward or rightward).
Excessive spatter	Voltage too high.
Irregular weld shape	Incorrect gun angle or changes in electrode extension.
Porosity	Insufficient shielding gas or heavily rusted plate.
Poor wire feeding	Wrong voltage or faulty contact tip or swarf in liner, wrong size or type of feed roll.

Table 6.3 *Typical welding currents for various wire diameters.*

Welding position	Wire dia. mm (in.)	Current (A)	Voltage
Self-shielded wires, DCEP			
All positions	1.6 (1/16)	150–450	21–40
Flat, HV	2.4 (3/32)	250–550	26–40
Rutile flux core, DCEP, Ar/CO_2 shielding			
All positions	1.2 (3/64)	150–290	22–30
	1.4 (0.055)	170–310	22–31
	1.6 (1/16)	190–330	28–32
Rutile flux core, DCEP, CO_2 shielding			
Flat, HV	1.6 (1/16)	200–400	23–32
	2.0 (5/64)	230–500	24–35
	2.4 (3/32)	250–600	26–38
Basic flux core, DCEN, CO_2 shielding			
All positions	1.2 (3/64)	100–250	18–26
Flat, HV	1.6 (1/16)	140–400	18–32
Flat	2.0 (5/64)	200–450	28–34
Flat	2.4 (3/32)	300–525	28–34
Metal core, DCEN, Ar/CO_2 shielding			
All positions	1.2 (3/64)	100–280	17–30
All positions	1.4 (0.055)	100–330	18–30
Flat, HV	1.6 (1/16)	150–400	18–30
Flat	2.4 (3/32)	350–550	30–36
Self-shielding, DCEP			
All positions	1.6 (1/16)	150–450	21–40
Flat	2.4 (3/32)	250–550	26–40

Voltage

As the arc voltage is reduced penetration is increased but lengthening the arc by increasing voltage raises the risk of porosity and undercut. When working in the dip transfer range at comparatively low currents, e.g., when positional welding, there is a risk of inadequate sidewall fusion and the arc voltage should be kept as high as is practicable.

Welding speed

Welding speed has an unexpected effect on penetration as the following example illustrates. When using a 1.6 mm (1/16 in.) metal-cored wire at 350 amp increasing the welding speed from 300 mm/min (12 ipm) to 600 mm/min (24 ipm) approximately doubles the penetration beyond the root of the fillet. A further increase of 300 mm/min (12 ipm) will result in a decrease in penetration as would the use of very low speeds. This behaviour is caused by whether the arc impinges directly on the plate or on the molten pool which screens the plate from the arc. Low welding speeds are avoided when impact properties are important as is the case with manual metal-arc welding.

Electrode extension

The increased sensitivity of cored wires to the I^2R resistance heating effect is made use of in two ways. Increasing the electrode extension but not changing the wire feed rate causes the current to drop and reduce heat input. If the wire feed rate is increased to restore the current to its original level, however, the deposition rate rises, an effect made use of in hardfacing applications. This is illustrated in Fig. 6.4.

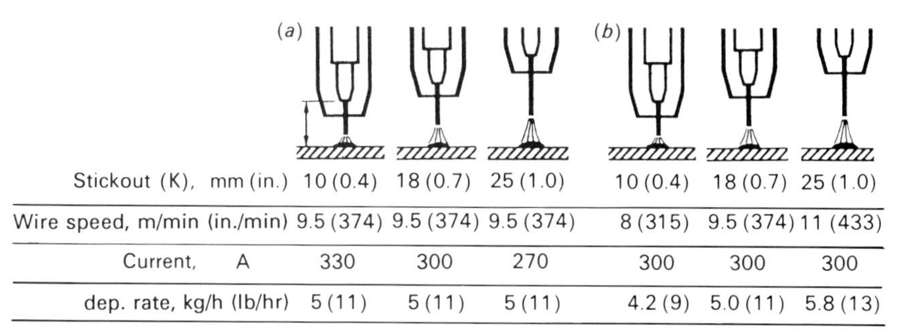

	(a)			(b)		
Stickout (K), mm (in.)	10 (0.4)	18 (0.7)	25 (1.0)	10 (0.4)	18 (0.7)	25 (1.0)
Wire speed, m/min (in./min)	9.5 (374)	9.5 (374)	9.5 (374)	8 (315)	9.5 (374)	11 (433)
Current, A	330	300	270	300	300	300
dep. rate, kg/h (lb/hr)	5 (11)	5 (11)	5 (11)	4.2 (9)	5.0 (11)	5.8 (13)

Fig. 6.4 *Effect of electrode extension (stickout) on* (a) *welding current and* (b) *deposition rate.*

Welding gun gas nozzles

Welding gun gas nozzles are normally 15 mm (5⁄8 in.) diameter and extend approximately 6 mm (1⁄4 in.) beyond the contact tip. When welding in confined joints a short nozzle may be used which is level with or 5–6 mm (3⁄16–1⁄4 in.) behind the end of the contact tip to permit satisfactory access. Adequate gas protection is then provided by the retaining effect of the edge preparation. It is recommended that nozzles of two or three different lengths should be available for use as needed on root passes and subsequent filling runs.

Gun angle and manipulation

There is a significant effect of gun angle on slag control and weld deposit profile. The appropriate gun orientations for standing fillet and butt joints are illustrated in Fig. 6.5, and for both types of joint rightward (backhand) or leftward (forehand) operation can be adopted. The rightward method which provides the best penetration, as the arc force keeps the slag from running in front of the liquid pool, generally gives a slightly convex weld bead. This method provides the maximum single-pass throat thickness, up to 8mm (5⁄16 in.) with 1.6 mm (1⁄16 in.) wires for standing fillet welds. Leftward operation is advantageous for thinner plate where deep penetration is not required but low heat input is needed to avoid distortion. A sufficiently high welding speed prevents the molten slag running ahead of the liquid pool. A further advantage is a

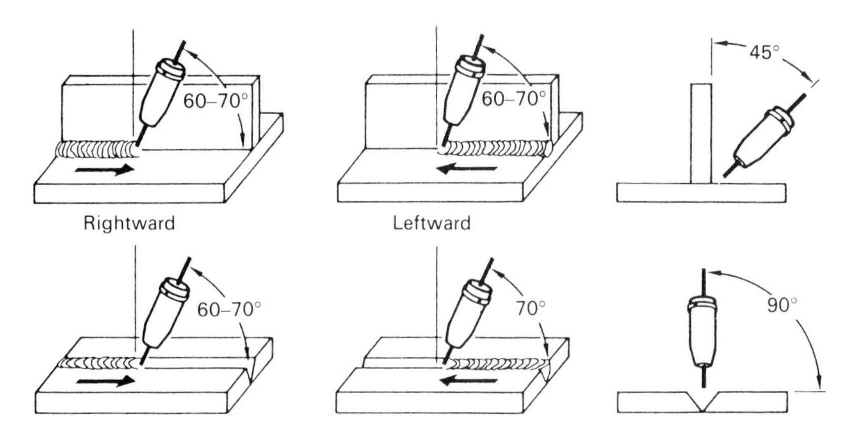

Fig. 6.5 *Manipulation of gun in cored-wire welding in the flat and HV positions.*

flatter weld bead. For standing fillet welds the gun should be directed slightly on to the horizontal plate to obtain equal leg lengths and avoid sagging.

Positional welding

Positional welding in butt and fillet welds is shown in Fig. 6.6 and Fig. 6.7. Rutile wires can only be used for root passes in butt welds with ceramic backing. Metal-cored and basic flux-cored wires provide good to excellent root penetration in butt welds with or without ceramic backing strips. A feather edge preparation can be adopted for 'V' joints when the vertical-down method is applied. Weaved runs are possible, as shown, but for optimum mechanical properties the split weave or stringer bead method is preferred. This applies to both vertical butt joints and multi-pass fillets. Vertical-down operation is generally adopted for thinner plates. The lower heat input, with thinner beads gives good Charpy impact values but this tends to cause higher tensile and yield strengths which may not always be desirable.

Small diameter rutile wires of the all-position type are generally preferred for positional welding because of their excellent spray transfer characteristics. Metal-cored and basic types can also be used in the dip-transfer mode but electronically controlled transfer, possible with the latest developments in power sources, considerably enhances the positional welding capability of these wires.

Plate preparation

The superior sidewall fusion obtained particularly from metal cored wires often allows the angle of the edge preparation to be reduced so saving weld metal. A 'V' butt joint would normally require a 60°

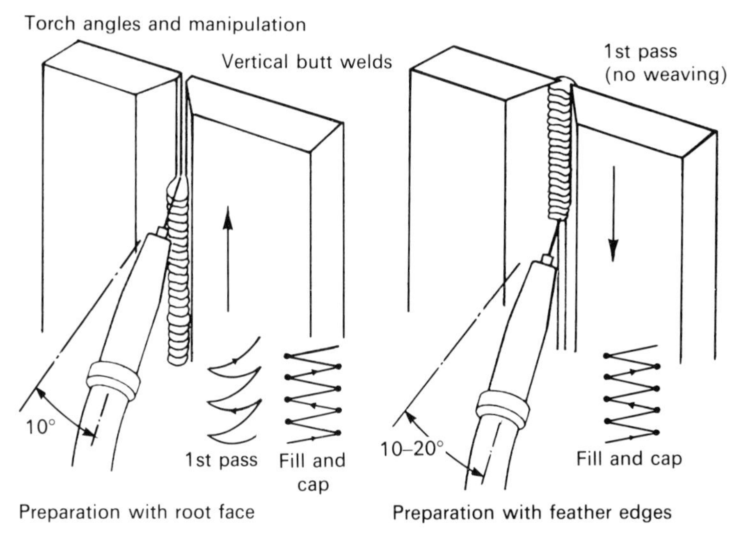

Fig. 6.6 *Manipulation of gun in cored-wire welding of vertical butt joints.*

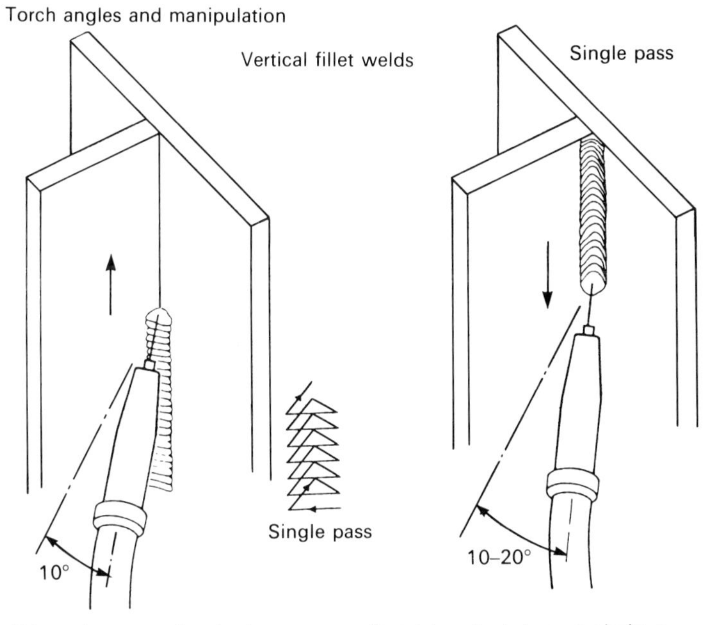

Triangular weave for single pass fillets. If necessary subsequent weld runs should be deposited using technique similar to that for filling vertical butt jointc.

Restrict vertical-down technique to thin plate or leg lengths of 6 mm (0.24 in.) maximum. May be used for first pass of multi pass joints.

Fig. 6.7 *Manipulation of gun in cored-wire welding of vertical fillets.*

included angle for manual welding but with cored wire this can be reduced to 45°.

Process assessment

Cored wires offer high deposition rates and give good penetration. Their use in preference to solid wire in MIG welding is usually justified on economic grounds. The presence of a slag permits positional welding when this would have been impossible with solid wire. Weld metal quality is good with satisfactory low-temperature properties and the process is not as sensitive to rust and scale as other processes. The self-shielding wires are more resistant to the effects of draughts than covered electrodes. Modern cored wires give low hydrogen weld metal. The range of wire compositions available is extensive and the manufacturing process for wires so flexible that special compositions or small quantities of wires can be obtained readily. Widely used in shipyards, for structural engineering and earth-moving plant, the process is being used increasingly with smaller diameter wires for engineering components. Cored wires are used occasionally in submerged-arc and electro-slag welding.

Chapter 7

Submerged-arc and electro-slag processes

Submerged-arc welding is so named because the arc between the electrode wire and the workpiece is submerged under a layer of powdered flux delivered in front of the electrode from a hopper. It was the first really successful machine method for arc welding and was developed in the 1930s. Any number of electrodes up to three is commonly used and the process is capable of welding thick material with high productivity in the H–V or gravity-welding position.

Electroslag welding was developed from the submerged-arc process by The Paton Institute in the USSR. The intention had been to develop vertical submerged-arc for the purpose of making vertical butt joints in a plate girder bridge over the river Dnieper. The powdered flux which melts to form a deep bath and the molten pool under this cover are held in position by water-cooled copper dams on each side of the joint. There is, however, an important distinction from submerged-arc welding in that there is no arc in the process. Molten fluxes are electrically conductive and current passes from the electrode wire to the molten pool without arcing but creating heat by electrical resistance.

Both submerged-arc and electroslag have a number of variants and both have specialised applications in the welding of thick plate; submerged-arc mainly for H–V or gravity-position welding and electroslag for vertical-up welding. Work has been done to develop both processes for other metals than carbon and stainless steels. Although these have met with some success for copper-base alloys, and also to a lesser extent in the past for titanium and aluminium alloys the overwhelming use of the processes is for welding structural and pressure vessel steels. The utilisation of electroslag on which much work was done in the 1960s and 1970s has declined in the 1980s. This is a consequence of improvements in the competing narrow gap submerged-arc process and of the intractable problems of heat affected zone properties in electroslag caused by the prolonged thermal cycle which normally makes post-weld heat treatment such as normalising necessary.

Submerged-arc

Principles

The submerged-arc process employs a continuous bare electrode wire, usually in solid form although cored electrodes are sometimes used for surfacing, and a blanket of powdered flux to shield the arc and weld area (see Fig. 7.1). The process exists in two forms, usually for automatic or machine welding but occasionally for semi-automatic welding. The automatic version is illustrated in Fig. 7.2. Because the wire-feed drive can be incorporated in the welding head close to the arc it is possible to feed quite large diameter wires. The semi-automatic version resembles MIG in that the wire drive is remote from the welding head and the wire is a relatively small diameter to allow flexibility in using the process. Flux is dispensed from a small trumpet-shaped hopper actually fixed to the welding gun. In both versions the arc is covered by the flux

Fig. 7.1 *Principle of submerged arc welding.*

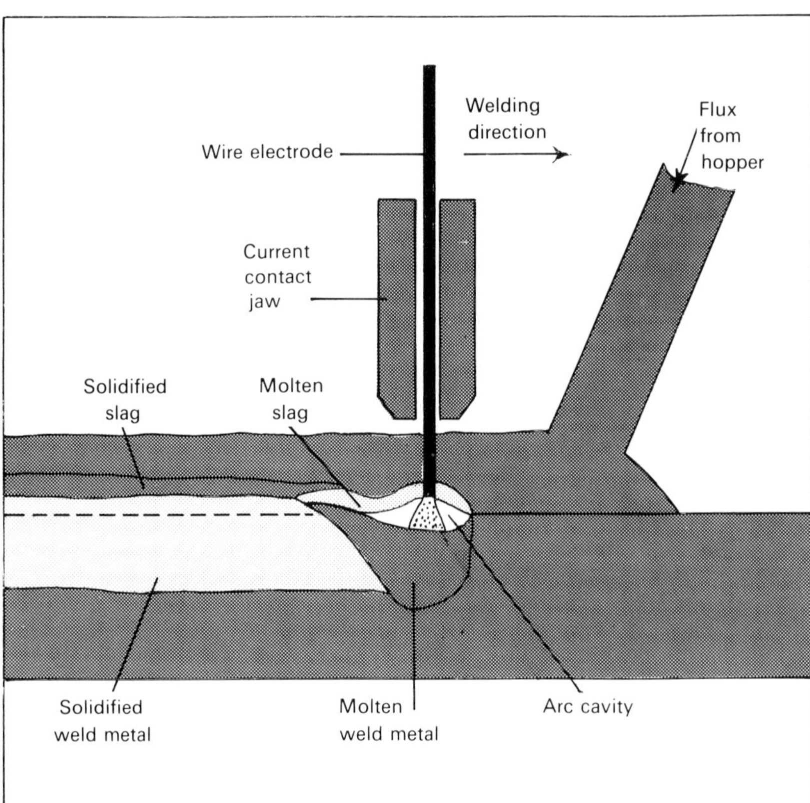

Fig. 7.2 *A tractor-mounted submerged-arc unit making a fillet weld.*

so it is impossible to view the molten pool during welding. With the automatic version it is possible to have several electrode wires, frequently two but generally no more than three, which are usually arranged in line one behind the other (tandem welding). It may be necessary to place individual weld beads separately for multipass welding, filling wide grooves or surfacing and the wires are then spaced apart laterally as well as longitudinally. If wires are close together they will feed into the same weld pool but when 60–120 mm (2½–5 in.) apart each wire forms a separate pool which can be helpful to mechanical properties and weld cracking. A form of the process used for cladding employs a strip electrode up to 130 mm (5 in.) wide and 0.5 mm (0.020) thickness. Because low penetration and dilution is desired in surfacing the DCEN polarity is used.

The function of the flux is much the same as in manual metal arc welding, i.e., to protect the arc from the atmosphere, to bring about desirable chemical change to the weld metal, to provide alloying elements and to control the shape of the deposited metal. A proportion of the flux melts covering both molten and solidified weld metal. Unmelted flux can often be recovered by a vacuum system and provided it is not contaminated may be used again. The melted and resolidified flux should release itself from the weld bead when this cools to a few hundred degrees Centigrade a short distance behind the welding head.

Because of the way the arc is submerged beneath the flux extremely high welding currents may be used without causing spatter or entrainment of air. As a result the arc is deeply penetrating and weld beads are frequently highly diluted with parent metal. This is of particular significance when considering such properties as notch toughness. Welding speeds are high and the excess metal when using a close square butt or small V edge preparation is minimised. Most welding is carried out in a 1000 amp range from 500–1500 amps but currents as low as 150 amps or as high as 4000 amps can be employed. The large slowly freezing molten pools resulting from the highest welding currents are liable to suffer from cracking while they and the HAZ may have low notch toughness. For this reason the use of very high currents is exceptional.

Both DCEN and DCEP as well as AC are used. The electrode positive polarity gives the deepest penetration and best shape as well as being the easiest on which to strike the arc. Deposition rates are highest with electrode negative and lowest with electrode positive with AC in between. The use of high welding currents makes arc-blow effects more noticeable. Earth connections are often made to both ends of a workpiece to provide stable conditions when working from one end of a workpiece to the other.

Multiple-wire techniques

Mention has been made of the possibility of using two of more wires (see Fig. 7.3). There are many ways in which this can be achieved, the simplest being to feed two wires through electrically-connected contact tubes connected to the same power source. This is known as 'two-wire parallel' welding and allows an increase in deposition rate of 100 per cent. It is more usual, however, for each wire to be supplied with current separately which allows various permutations of polarity, DC or AC. Penetration is greatest with DCEP and it is common for this to be used on the lead electrode in both twin- and triple-wire welding. The second wire can then be either DCEN or AC, with both of which current supplies penetration tends to be less than on the positive polarity. It is usual for the second electrode to point forward slightly as this provides a flatter bead surface. With a three-electrode system the third or trailing electrode points even further forward. Two-wire AC systems are also used and with three wires either DECP-AC-AC or AC-AC-AC. Arcs with the same polarity pull together if they are close enough.

Fig.7.3 *Two gantry-mounted three wire submerged-arc units in a shipyard panel line.*

AC arcs are less susceptible to such arc blow but it may still occur if both current supplies are of the same phase. The Scott type three to two phase transformer can be used as this gives a 90° difference in phase between the two outputs.

Submerged-arc variants

Iron powder additions

The deposition rate can be increased without raising the welding current by adding iron powder to the joint from a separate dispenser ahead of the flux delivery tube, Fig. 7.4, or by adding iron powder around the eletrode to which it adheres because of the induced magnetic field (see Fig. 7.5). The iron powder is melted into the molten pool increasing its volume, reducing its temperature and decreasing the dilution of the parent material. The main advantage of using metal powders is the increase which is possible in deposition rate and hence the improved productivity. Welding conditions must be controlled precisely if attempts are to be made to alter the weld composition and hence the mechanical properties of the weld by the use of powders with a different composition from the plate or wire.

Hot wire and I^2R techniques

Other ways to increase deposition rate apart from using iron powder are to employ hot wire or I^2R effects. In the hot-wire method a separate filler wire is fed into the molten pool and is connected to its own power source. The melting rate is substantially increased by the resistance heating of the filler wire but the current drawn is limited so that the wire is preheated to as much as red heat but no arcing occurs. Hot-wire techniques are also used with TIG welding, see page 148. Another way to use the resistance heating effect to increase deposition rate is to increase the electrode extension by fitting an insulated extension piece to the guide tube. This makes use of what is often called the I^2R effect and was discussed in Chapter 6.

Narrow gap techniques

With V or J edge preparations the metal which must be added to fill the groove increases progressively as plate thickness increases. For example,

Fig. 7.4 *The 'Bulkweld' process for iron powder additions.*

Fig. 7.5 *The 'Oerlikon' method for iron powder additions.*

the cross-sectional area of a single V groove is $t^2 \tan \frac{1}{2}\propto$ where t is the plate thickness and $\propto$ is the angle of the groove. As the need for welds in thicker material has developed in the last two decades so has the incentive to find a way of reducing the angle of the preparation. The first such attempts were made with MIG welding but these were not wholly successful although after much development work on equipment and power sources narrow gap MIG is now in use in certain countries. Submerged-arc welding had the potential for narrow gap welding but slag removal and inadequate sidewall fusion were initial problems. Both these difficulties have been overcome and narrow gap submerged-arc welding is now used frequently in the boiler and pressure vessel industries for welding thicknesses over 100 mm (see Fig. 7.6).

The maximum advantage from narrow gap welding is obtained with a single pass per layer. This is, however, the most critical technique and

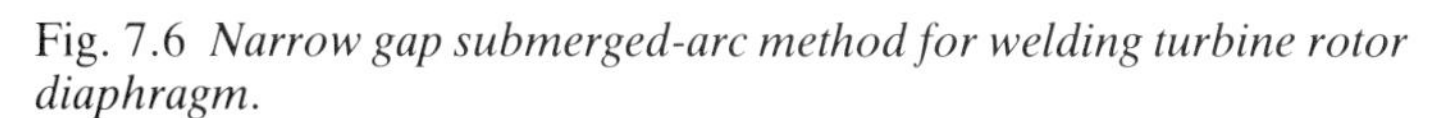

Fig. 7.6 *Narrow gap submerged-arc method for welding turbine rotor diaphragm.*

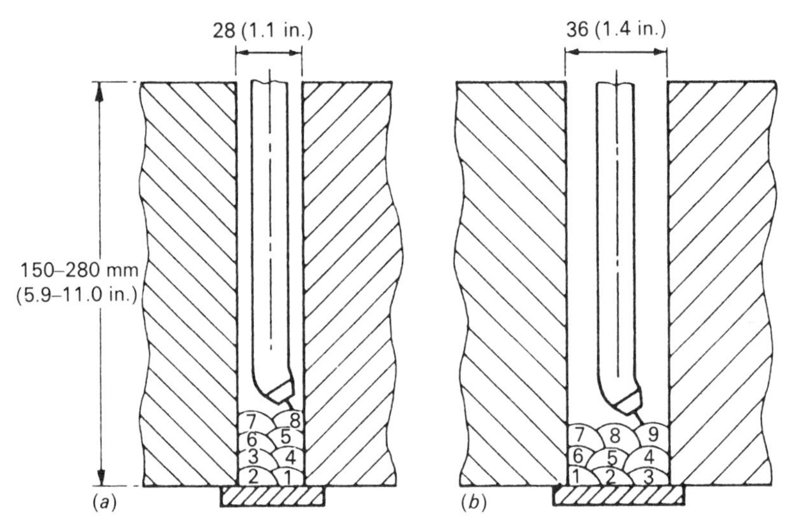

Fig. 7.7 *Pattern for weld bead deposition in narrow gap submerged arc: (*a*) two beads per run and (*b*) three beads per run.*

fewer problems with slag removal and lack of fusion are experienced with a two pass per layer technique in which alternate passes fuse first one side and then the opposite side of the groove (see Fig. 7.7). This increases the welding time to half as much again but this is not a great penalty to pay for improved reliability and defect free welds. The wire diameter employed is related to the groove width but is usually 3–4 mm (1–5⁄32 in.) diameter. Within the plate thickness range 100–250 mm (4–10 in.) the groove width varies between 14 and 20 mm, (0.55–0.8 in.).

The flux used in narrow-gap welding is of crucial importance as it must provide a stable arc, have excellent detachability and provide acceptable notch toughness in weld metal. With many fluxes detachability improves as silica content increases but this increases the silicon content of the weld metal which can limit notch toughness. Selection of the flux for narrow gap submerged-arc is therefore a matter of compromise and although there is scope for further development a number of cases are on record of successful flux development. Power sources for narrow-gap welding must provide an arc which is stable and not subject to arc blow. AC is therefore preferred and square wave AC with which current reversals do not cause extinction of the arc is desirable. Current requirements are up to 650 amps and voltages 27–35.

Perhaps the most critical advantage of narrow gap submerged arc techniques is the opportunity they present to improve substantially the output per man and per station as compared with conventional submerged arc welding. A fully mechanised technique will give welds almost free from defects but more importantly from the productivity point of view the reduced heat input can make unnecessary the inter-run cooling periods, often required with conventional submerged arc when welding to demanding notch toughness requirements.

Equipment

Power sources and feed units

Power sources for submerged-arc welding must be of rugged construction and designed for a 100 per cent duty cycle. Because of the high welding currents required it is fairly common to connect two or more similar power sources in parallel to provide the desired capacity. If this is done with AC plant the outputs must be in phase. To ensure reliable reignition of the arc following current reversal on AC the open circuit voltage should be at least 80 volts. DC power sources are either constant current (drooping characteristic) used with a voltage-sensitive arc length control welding bead or constant voltage (flat characteristic) type used with a constant speed wire feed. With the 'constant current-arc length control' system the wire feed rate is adjusted according to the arc voltage so as to keep a constant arc length.

The 'constant voltage-constant wire feed speed' arrangement is a DC single-wire system the same as that described in Chapter 5 page 83 and makes use of the self-adjustment effect. It is a good system to use on relatively thin sheet metal applications requiring high-speed welding, changes of direction or intermittent welding. The high short-circuit current available with this arrangement makes arc striking easy. Some welding heads are available in which the electrode feed and tractor carrying the welding head are driven from the same motor. In this way the amount of deposited metal per unit length of joint is a preset constant. Adjustable gearing allows the ratio of metal deposited to length of joint to be altered.

With the welding of thicker plate material the 'constant current-voltage control' option is usual, the larger diameter wires employed being less susceptible to the self-adjusting effect. Except when wires are being used in parallel it is usual for each wire to have its own drive and voltage control unit. Welding heads are frequently designed to allow the electrode to be retracted automatically to strike the arc. In the absence of this facility it may be necessary to place a ball of wire wool between the electrode and the work to provide a temporary path for the current and allow the arc to strike. Welding heads may also have a crater-filling facility in which forward motion of the head is halted but wire is fed for a short period of time to fill the crater.

Tractors and manipulators

With longitudinal welds it is usual for the welding head to be moved over the stationary work. Single-wire units can be mounted on small self-propelled tractors which often run on a guiding track held by temporary fixings to the work. Tractors can also be steered along a joint, without a track, either manually or with guiding wheel or other seam tracking device. If a gravity position fillet is being welded, e.g., in a beam, the work itself provides the guidance. Overhead gantries from which single or multiple welding heads are slung are popular as they offer minimum encumbrance in the working area. They are employed for making the twin fillet welds between the flange and web of plate girders or on panel lines for attaching stiffeners to plate in shipyards. With this as with other twin fillet types of application it is normal for one welding head to lead the other by 75–100 mm (3–4 in.) as this helps to avoid the trapping of gases and porosity as well as reducing the risk of cracking.

In pressure vessel shops the column and boom is indispensable. Longitudinal welds in vessels are often made by having the boiler drum on rollers and positioning the joint at the 12 o'clock position. The boom, along which the submerged-arc equipment travels, is arranged to be in line with the joint. Circumferential welds are made by positioning the welding head with the column and boom and revolving the drum on rollers under the head. Another standard piece of manipulating plant often used with submerged-arc is the rotating turntable, the axis of which can be set to angles between horizontal and vertical. These turntables are made in sizes ranging from those carrying a few kilograms (lbs) to capacities of many tonnes. Whatever piece of plant is used to move welding head or work its controls will be interlocked with those on the welding head and power source.

Welds on curved surfaces

When making a longitudinal weld in a circular workpiece or pipe with a diameter less than about 1 m or 36 ins the flux may tend to fall away and on such diameters it would be generally preferred to use a process such as MIG. Somewhat similar considerations apply when making circumferential welds and there is a workpiece diameter below which flux will not stay in position. Another consideration when welding circumferential joints in drums is that to provide the best shaped weld bead solidification must take place at the 12 o'clock position when welding outside or 6 o'clock when welding inside. This is achieved by displacing the welding head so that as the molten pool is created it is carried towards the 12 o'clock position where it solidifies. Too much displacement means that the molten pool will become wide and concave but too little results in solidification beyond the 12 o'clock position and narrow peaky weld beads. When welding inside a drum the above comments must be reversed.

Consumables

Wires

While it is possible to make alloy additions through the flux in submerged-arc welding in the same way as with manual metal-arc, more accurate control of the weld metal composition is usually achieved by employing a range of wires of varied compositions. For use on carbon and carbon manganese steels there are four common wires differing mainly in manganese content which varies from 0.5–2.0 per cent. Higher strength deposits are to be expected from the higher manganese wires but this performance is greatly affected by the flux which is used. Where a higher strength is required high manganese wires with an additional 0.5 per cent Mo are used and where good impact properties are especially required carbon manganese wires with either 1.0 per cent Ni or 2½ per cent Ni are available. With creep-resisting and high-strength low-alloy steels Ni/Cr/Mo wires are used. For stainless steels, wires of the required composition are usually employed. The flux is compounded to ensure that alloying elements are not lost but it is not a vehicle for introducing alloying elements. As mentioned above it is possible to make additions through the flux but unlike manual metal-arc where the ratio of wire to flux melted is substantially constant regardless of current and voltage, with submerged-arc this ratio can vary according to the welding voltage. It is therefore more difficult to guarantee the ultimate analysis of the deposit. Cored wires containing both chromium and molybdenum are used for hardsurfacing applications. Other cored wires can be used instead of solid wires for special purposes, where for example a composition is required in the weld which cannot be produced by standard wires.

Wires for submerged-arc are usually copper-coated and come in sizes from 2.0 mm to 6.0 mm (5⁄65–¼ in.), the most popular part of the range being 2.4–4.0 mm (3⁄32–5⁄32 in.). They are covered in BS 4165 and AWS A5.23.

Fluxes

Fluxes for submerged-arc welding are made in several ways. The original and simplest way is to fuse the minerals together and comminute the product. Fused fluxes are made from the minerals lime (CaO) fluorspar (CaF_2), magnesia (MgO), quartz or silica (SiO_2), alumina (Al_20_3) and manganese oxides MnO and MnO_2. According to the proportion of silica and alumina to lime and magnesia (the basicity index) the flux may be acid, neutral or semi-basic. Fused fluxes are homogeneous and non-hygroscopic. The fines can be removed without affecting the composition and they are therefore safe to re-cycle.

Unfortunately it is difficult to incorporate certain deoxidisers and ferroalloys into fused fluxes.

Another way of preparing submerged-arc fluxes, called agglomeration, is to dry mix the powdered minerals and then bond them with potassium or sodium silicate. Alternatively, the dry constituents may be bonded by fritting them together with heat. The fluxes are then described as sintered. Using these techniques deoxidisers, alloying elements and minerals subject to decomposition by heating can all be incorporated in the flux. Agglomerated fluxes are more susceptible to the pick-up of moisture and can suffer from segregation as the composition of the fines may be different from the bulk of the flux.

During the welding process the molten flux reacts with the wire and molten pool and elements, notably manganese and silicon, pass from the slag to the weld metal. With high silica fluxes the increase in silicon content of the weld metal over wire and plate can be appreciable. In a single-pass weld this may not matter particularly but with multipass welds the silicon builds up because of dilution effects. This can be undesirable from the cracking and weld toughness point of view and it is normal therefore to use low-silica fluxes for multipass welding.

Particles size and distribution affects the performance of a flux. Coarse particles with a small size range give a permeable cover which is desirable as it allows gases to escape but the particle size must be reduced as the welding current increases to contain the arc. Too high a current for a given particle size causes the bead to have ragged uneven edges. The depth of the flux over the joint also affects the appearance of the bead. If it is too shallow there will be flashing, sputtering and a rough pock-marked surface to the weld. When the flux cover is too deep gases may be prevented from escaping and can cause porosity.

The properties of the weld metal are influenced by the acidity or basicity of the flux. Acid fluxes high in silica tend to give weld metal high in oxygen, basic fluxes give clean weld metal low in inclusions. In an attempt to quantify these effects various 'basicity indices' have been devised of which one well-known index is that devised by Tuliani, Boniszewski and Eaton:

$$\frac{CaO + CaF_2 + MgO + K_2O + Na_2O + \tfrac{1}{2}(MnO + FeO)}{SiO_2 + \tfrac{1}{2}(Al_2O_3 + TiO_2 + ZrO_2)}$$

Where the index is 1.00 the flux is neutral, if it exceeds 1.00 it is basic and if less than 1.00 it is acid.

Welding techniques

Submerged-arc welding produces high-quality weld metal at a high deposition rate, with deep penetration if required. It is a process widely used in structural engineering, shipbuilding, offshore structures and process plant manufacture. Although its main use is on material of considerable thickness it may also be used on sheet metal down to about 5.0 mm (0.2 in.) thickness providing there is suitable backing. With unsupported welds the minimum thickness is about 8 mm (5/16 in.). The process is applicable to carbon, carbon manganese, low and high alloy steels and stainless steels. Of the non-ferrous metals only copper alloys can be welded satisfactorily. Its utilisation at around 8 per cent of the total amount of weld metal deposited has been remarkably consistent over the years.

It is a process which can only be used to make butt and fillet welds in the flat or horizontal-vertical positions. Horizontal butt welds as for example in storage tanks can be made with special devices to support the flux. Because of the relatively large molten pools the process is quite sensitive to the inclination of the workpiece. A few degrees will make a difference to penetration and bead shape.

As with processes such as MIG it is the welding current which determines penetration and melting rate. At the extremes an extra high current will give a 'digging' arc and a narrow weld bead while arc instability results when the current is excessively low. With all other conditions the same increase in arc voltage (arc length) will give a flatter wider weld and will increase the consumption of flux together with a concomitant increase in the pick-up of elements, e.g., Mn and Si, from the flux. Carried to extremes the lengthening of the arc will give concave beads, increase undercut and impair slag removal. Welding speed directly affects the amount of metal deposited per unit length and hence weld reinforcment.

The diameter of the electrode wire has an effect on deposition rate, smaller diameter wires giving greater deposition rates for the same welding current. This effect was mentioned in the discussion of cored-wire welding (page 110) and has to do with the electrical resistance of the electrode extension. Because an increase in electrode extension means a greater resistance and voltage drop in the extended wire there is an associated drop in arc voltage. This should be corrected if the electrode extension is increased. Deposition rates can be increased by as much as 50 per cent at the same current by lengthening the electrode extension but penetration will fall noticeably.

Although very large welds can be made in a single pass it is more usual to employ multipass welding where, because of grain refinement in reheated weld metal, the toughness properties are improved. There is also a reduced risk of centre line hot cracking.

Because the process can be deeply penetrating or have large molten

Fig. 7.8 *Magnetic clamps and backing strip for butt welding plate in a panel line.*

pools attention must be given to backing systems. There are a variety of methods such as fusible or permanent steel backing bars, a MIG or MMA welded root, ceramic or flux backing. On large installations such as panel lines in shipyards elaborate backing assemblies are often used in which the plates are clamped over channels filled with flux (see Fig. 7.8).

Process assessment

The submerged-arc process exists in several forms, semi-automatic, automatic with single electrode or multiple electrodes and special techniques such as hot wire and iron powder addition. It produces quality welds in steels at high speed particularly in heavy plate although it can be used down to sheet metal 5.0 mm (0.20 in.) thick, but here it is in competition with flux-cored and solid wire MIG. The process is also used for surfacing when multiple or strip electrodes are employed. The main fields of application are in shipbuilding, structural engineering and process plant industries. Its limitatons are the need to support the flux, and that welding must be done in the H–V or flat position.

Electro-slag welding

Principles

There are three variants of electro-slag welding according to the type of electrode employed, wire, consumable guide or plate, all of which can only be used for vertical welding with a deviation from the vertical of 15 degrees maximum.

Wire electrode electroslag

The equipment comprises a carriage which is elevated by a rack mechanism on a column placed alongside the joint (see Fig. 7.9 and Fig 7.10). On the carriage there are usually between one and three guide and contact tubes through which the electrode wires are fed changing direction from the horizontal to the vertical-down direction. One motor may feed all the wires or there may be a separate motor for each wire. The reels of electrode filler wire are also generally mounted on the carriage. Water-cooled sliding shoes are mounted on the carriage, the one on the nearside of the carriage and rack mechanism being supported from just below the wire feed mechanism while the shoe on the far side is hung from a bar which passes through the joint above the wire feeders. Both shoes have pivoting devices and screw clamps so that the shoes can be pressed firmly on each face of the joint. The shoes must retain the molten metal and flux so the surfaces of the plates on which they bear are usually lightly ground to ensure smooth sliding. Shoes are frequently made to allow some flexibility by being in three pieces and articulated. The carriage has a variable-speed drive by which it is hoisted

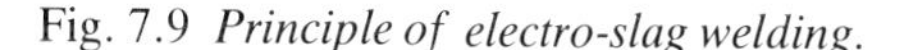
Fig. 7.9 *Principle of electro-slag welding.*

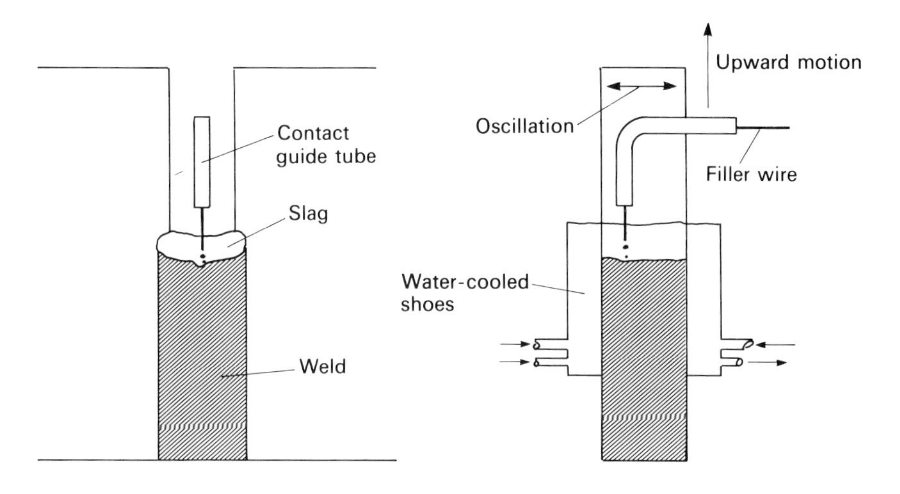

Fig. 7.10 *Electro-slag welding a boiler drum.*

up the rack on the column and an oscillating mechanism to move the electrodes to and fro through the thickness of the joint so as to promote uniform melting.

Welding is started on a U-shaped extension piece under a layer of powdered flux. Initially the wire arcs to the work melting the workpiece and the flux and a layer of molten flux quickly develops over the molten pool. More flux is then added and once the molten slag bath has become about 50 mm (2 in.) deep the process changes with the current being conducted through the slag without arcing. The resistance heating which results produces a hot molten slag bath which melts the wire and the joint faces. The upward motion of the carriage then begins and flux additions are now reduced to keep a constant level of the slag bath, replacing the slag which is lost in forming a thin layer between the shoes and the workpiece. Upward motion keeps pace with the rate at which the electrode wire melts off to fill the joint. A common electrode size is 3.2 mm (⅛ in.).

A single electrode can be used on metal up to 100 mm (4 in.) thickness and the wire can be centrally placed without oscillation for thicknesses up to about 50 mm (2 in.). Above this thickness the wire is oscillated in the joint gap with a dwell period at each end of the stroke to compensate for the heat lost to the water-cooled shoes. A second electrode is added when the workpiece thickness exceeds about 100 mm (4 in.) and a third at about 250 mm (10 in.) thickness. A 'constant voltage-constant wire feed system' is used with 300–500 amps per wire. Conditions such as traverse and end dwell are adjusted to give a weld with a cross section as near parallel as possible but generally the result is a barrel shape.

Although electroslag welding has been used on plates as thin as 20 mm (¾ in.) when welding speeds of up to 9 m/hr (30 ft/hr) were achieved it is now rare for the process to be used for plate less than 75 mm (3 in.) in thickness. Speeds of 1 m/hr (3 ft/hr) are normal in this range, higher speeds resulting in a tendency to centreline weld cracking.

The electrogas process, a MIG welding variant, is sometimes used for similar applications and the two pieces of welding equipment have some features in common. The electrogas process does not normally submit the workpiece to such extended thermal cycles as electroslag.

Consumable guide electro-slag

The same principle of a molten slag bath is employed with this process but the equipment and the operation of that process is greatly simplified. It is an electroslag variant uniquely suitable for making short lengths of weld in material of 25–75 mm (1–3 in.) thickness (see Fig. 7.11).

One or more hollow rods or bars, the consumable guides, which may be covered in flux like heavy-gauge stick electrodes are arranged in the joint gap. The slag bath is established as for the wire electrode variant with the electrode being fed through the guide. The molten slag covers the bottom of the guide which along with the electrode wire is melted away to fill the joint.

For a short joint of say 300 mm (12 in.) the water-cooled shoes are clamped on each side and are not moved. With longer joints an additional pair of shoes are employed, the two pairs being leap-frogged up the joint. Joints up to 9 m (30 ft) in length can be made in this way but a full-length guide must be used. Guides are normally about 10 mm

Fig. 7.11 *Consumable guide welding a web in a heavy girder.*

(0.4 in.) diameter and 1 m (40 in.) long but special purpose-made guides which can be joined together can be used. Consumable guide welding avoids the complicated series of motions necessary with the wire variant, the only movement needed being the wire-feed which as for the wire-electrode method is fed at a constant speed.

Plate electrode electro-slag

The third variant of the process is simpler still but is less commonly used. No wire electrode is employed but instead one or more plates are suspended in the joint gap and are lowered slowly as they melt away.

This variant was developed for short joints in very thick plates, e.g., 1 m × 1 m (40 × 40 in.). Water-cooled shoes are not used, the joint being closed instead by permanent 'backing bars' welded in position. A variation of the plate electrode process has been used for surfacing.

Fluxes for electro-slag

The fluxes employed are generally similar to those used for submerged-arc but are higher in fluorides. Because the slag bath is in contact with the molten pool for much longer than with submerged-arc, reactions between the slag and metal proceed further. In particular FeO may build up in the slag bath causing surface arcing. After a short length of weld equilibrium is established and weld metal composition remains stable. Additions of flux are usually made manually, the operator judging the depth of the slag bath by probing with a steel rod. Another possibility is to employ a cored wire but in this case the cored wire must not add an excess of flux. The flux so introduced should be less than the normal requirement, allowing the operator to make up the difference by occasional separate additions. Control of the weld metal composition is through the electrode wire, the wire guide and sometimes by the addition of metal powder.

Process assessment

Electroslag welding is capable of making single-pass welds with high productivity in steel of almost any thickness. Because of the shape of the weld there is little angular distortion and low transverse stress. Hydrogen cracking is not a problem and no preheating is required. The joint preparation being square edged is simple to produce. Applications which are all vertical or near vertical include heavy structural members, often involving forgings. The consumable guide method is particularly useful for joining the webs of members meeting in complex arrangements. Circumferential welds can be made but the technique for joining up with the start of the weld is demanding and other methods such as narrow gap submerged-arc or narrow gap MIG arc are now preferred. For vertical welding applications the electrogas process or mechanised MIG are serious competitors. The major limitations of the process are the difficulty of repairing defects although these tend to be rare and the problem of the low notch toughness of the HAZ because of the prolonged thermal cycle which may necessitate normalising. The process has been used for surfacing rolls but the equipment is capital intensive.

Chapter 8

TIG and plasma arc processes

Frequently used names for the process – GTA, gas tungsten arc; TIG, tungsten inert gas, (WIG in Germany).

In all the processes described so far, the arc has been struck from a consumable electrode to the workpiece and metal has been melted from the electrode, transferred across the arc gap and finally incorporated into the molten pool. Called GTA in the USA and TIG in many other countries, the process employs an electrode made from metal with a high melting-point, usually tungsten, which is not melted. Actually the tip of the electrode is fused but this molten end remains in place and none of it should find its way into the weld if the welding technique is satisfactory. This electrode and the molten pool is shielded from the atmosphere by a stream of inert gas which flows around the electrode

Fig. 8.1 *Principle of TIG welding.*

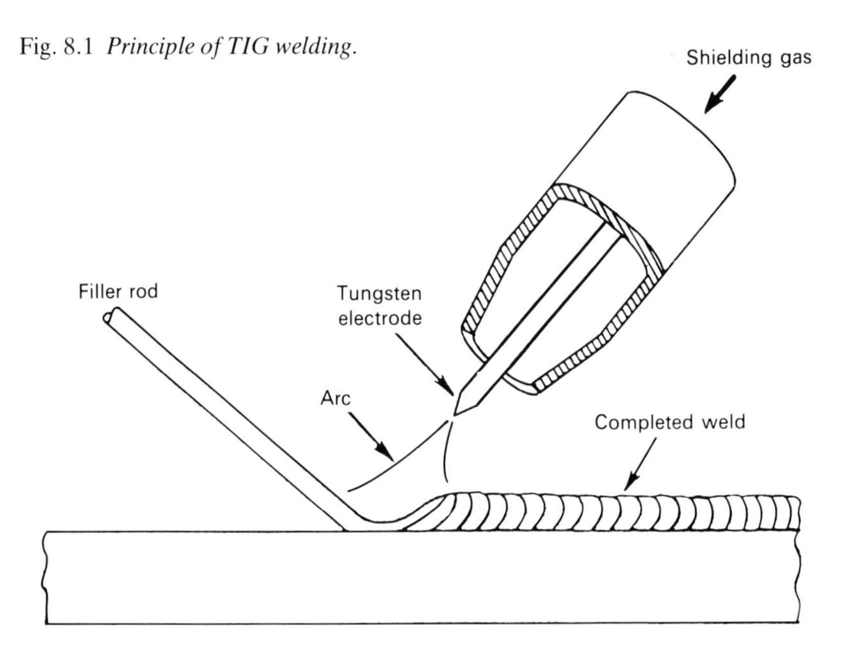

and is directed on to the workpiece by a nozzle which surrounds the electrode (see Fig. 8.1).

In TIG welding the primary function of the arc is to supply heat to melt the molten pool and any filler metal which may be necessary must be added separately. This is done by feeding bare filler wire mechanically or manually into the molten pool at its leading edge. A second function of the arc is to clean the surface of the molten pool and the immediately surrounding parent metal of surface oxide films and therefore no flux is required. The shielding gas must be inert with respect to the tungsten electrode and the choice is therefore more limited than with the MIG process.

The plasma arc process is a development of TIG in which the arc is forced to pass through a water-cooled metal collar. This has the effect of constricting the arc plasma so that the arc assumes a narrow columnar shape rather than the cone or bell shape of the TIG arc. The columnar arc is highly directional and is capable of penetrating deeply into the workpiece. The form of plasma arc just described is called a 'transferred' arc, one pole of the arc being the electrode and the other the work (see Fig. 8.2). If on the other hand the constricting collar is made the second pole of the arc there is no electrical connection to the work and therefore no arc to the work. However, a hot plasma flame emerges from the

Fig. 8.2 *Principle of plasma arc welding.*

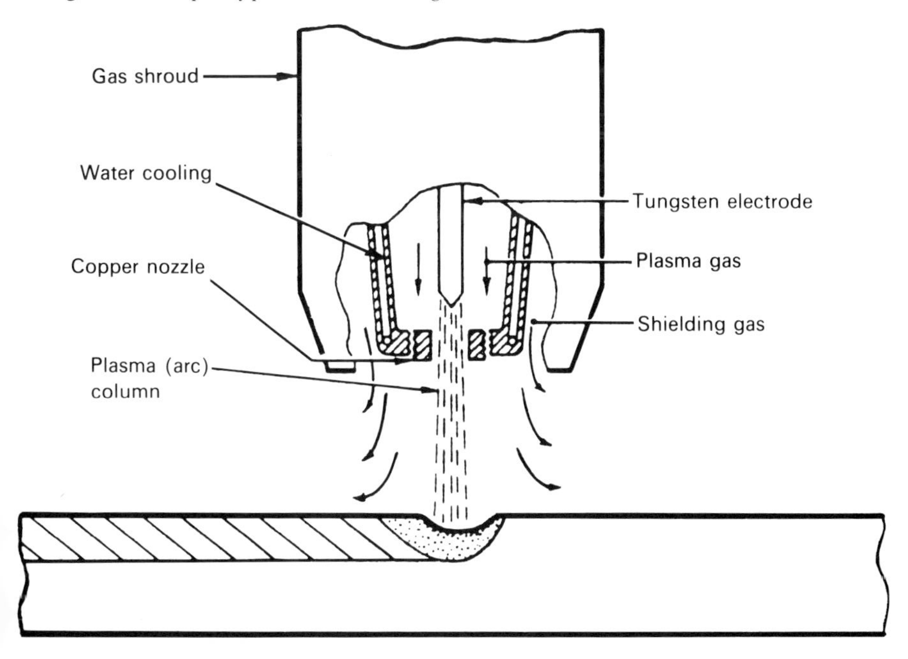

F

orifice in the torch which may be used for surfacing although it is generally not suitable for welding. Such a system is called a 'non-transferred' plasma arc. As with the torch of its sister process TIG, the plasma welding torch is primarily a device for supplying heat and if any filler metal is required it is added separately.

Both processes and their variants can be classed as precision welding methods. They can be used at low welding currents to weld thin material; are applicable to a wide range of materials, even reactive and refractory metals such as titanium and molybdenum; are clean to operate as no flux is used; can be precisely controlled because filler wire feed is independent of arc current and they are readily mechanised or automated. High-quality welds are possible. Although the main field of application is on thin material the process can be used to weld heavy plate particularly for root runs where the good control of penetration permits the making of quality unbacked welds. Variants of the process enable the economical welding of material of considerable thickness.

The TIG process

Principles

The torch for TIG welding is illustrated in Fig. 8.3. (Note that *The International Welding Thesaurus* describes the device for creating heat in the absence of a flame as a 'torch'. For gas welding where a flame is employed the term 'blowpipe' is used, except in the USA where 'torch' is preferred. The term 'gun' is used for a device through which filler wire

Fig. 8.3 *A TIG welding torch, cable cover detached from handle to show power, gas and control leads.*

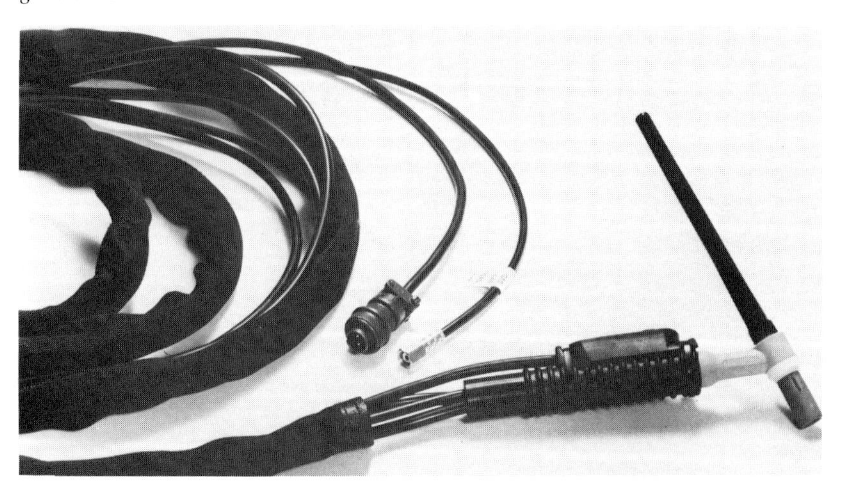

is fed; hence MIG guns but TIG and plasma torches.) The shielding gas is directed through a nozzle surrounding the electrode which is held in a collet fitted into the gas- or water-cooled body of the torch. This collet allows the electrode to be positioned so that it projects slightly beyond the end of the nozzle. The nozzle is usually air-cooled and made from refractory alumina ceramic.

Electrode material varies according to the type of current employed. For DC a thoriated tungsten or pure tungsten is used but the thoriated tungsten gives better arc striking and re-striking because of its higher electron emission. With welding currents less then about 50 A when electrodes of 1–1.6 mm (1/32–1/16 in.) are used the end of the electrode is ground to a 30° angle but the extreme point is taken off or it would be quickly melted off. Larger diameter electrodes are used at higher welding currents. At 250 A on a 3.2 mm (1/8 in.) diameter electrode the angle may be increased to 120°. When using DC the DCEN electrode negative polarity is almost invariably employed. DC must usually be chosen for welding alloy steels, stainless steel, titanium and copper (but not for copper alloys containing aluminium). The cleaning function of the arc does not take place on the DCEN polarity so metals forming refractory oxide surface films such as aluminium cannot be readily welded on this polarity.

For aluminium alloys the electrode positive polarity on which cleaning takes place would therefore appear desirable. In fact on this polarity less energy is liberated at the work and more at the electrode which therefore becomes overheated with a tendency to spit tungsten particles into the weld. Aluminium alloys are therefore welded using AC and the cleaning action takes place on the electrode positive half cycles. Where the correct welding current is used on the right diameter of electrode the end of the electrode becomes a stable molten ball. Zirconiated tungsten electrodes are often chosen for AC welding because the zirconia helps the electrode to maintain the desired stable end. Typical conditions for tungsten electrodes are given in Tables 8.1 and 8.2.

Table 8.1 *Electrode, current and gas nozzle sizes.*

Electrode D mm (in.)	DC thoria A	AC zirconia A	Pure tungsten A	Nozzle D mm (in.)
0.5 (0.020)	5–20	5–20	5–15	6.4–8.0 (1/4–5/16)
1.0 (0.020)	10–80	10–60	10–50	6.4–8.0 (1/4–5/16)
1.6 (1/16)	30–150	30–120	30–100	6.4–9.8 (1/4–3/8)
2.4 (3/32)	80–250	80–180	80–160	9.8–12.7 (3/8–1/2)
3.2 (1/8)	—	120–250	120–210	11.2–12.7 (7/16–1/2)

The AC is a balanced square wave.

Table 8.2 *Vertex angle for electrodes used on DC.*

Current A	Electrode angle
up to 20	30° not rounded
20–100	60–90° slightly round
100–200	90–120° slightly round

Arc initiation and maintenance

For many years AC power sources for TIG had sinusoidal current output. This made it necessary to provide a means for reigniting the arc after each reversal of current. This was either a high open circuit voltage, typically 100 V, or the provision of a continuous high-frequency high voltage. Modern square wave AC power sources make these provisions unnecessary but the high frequency voltage may still be necessary to ignite the arc. An alternative method of igniting the arc is to briefly touch the electrode on the work and the arc then strikes when the electrode is withdrawn. Touch starting carries the risk of tungsten being detached and contamination of the electrode by the workpiece when the short-circuit current surges through the point of contact. To avoid this the electrical circuit includes a device to limit the initial current surge and ideally the current should be increased gradually to its operating level over a period in the range of 0.1–10 sec (slope-up). The same device is used to reduce the current gradually when welding stops to allow time for the crater to fill (slope-down).

Once the electrode is hot the arc is readily ignited on the DCEN polarity because of thermionic emission from the electrode. While cold, however, and before thermionic emission can take place the arc can be ignited more readily on the DCEP polarity. Some welding sets therefore incorporate a starting device in which for a brief period, up to 20 msec, the electrode is made positive before being switched over to DCEN for welding. This device is particularly useful for larger electrodes over 3 mm (⅛ in.) diameter which take longer to reach emission temperature.

The TIG process relies entirely on the shielding gas to protect the hot electrode and molten pool and it is therefore essential that the flow of gas is initiated and allowed to stabilise up to 3 sec before the arc is struck (pre-flow). Equally it must be allowed to flow momentarily after the arc is extinguished (post-flow) to prevent oxidation of the electrode.

Equipment

Power sources

All power sources for TIG welding have a drooping or constant current output characteristic. DC power sources as used for welding alloy steels, stainless steels, etc., are usually of the rectifier type with stepped, thyristor, transistor control or inverters as described in Chapter 2. Unless the power source has been specifically designed for TIG welding, however, it is necessary to have an 'add-on' unit incorporating the arc striking and maintaining devices, the sequencing controls to allow slope-up and down, gas flow control and even a pulsing facility. These add-on units are made with varying degrees of sophistication from the basic necessities up to full sequencing and pulse control and are equally applicable to AC transformer and DC rectifier power sources.

Most custom-made power sources for TIG have an AC/DC output and are of the thyristor or inverter type. Although transistor-controlled sets are excellent for TIG welding they are more expensive and are therefore not commonly used on straightforward applications. The AC output is often square wave and the modern electronic power source makes it possible to alter the balance between the opposite half cycles in AC. Typically the balance may be varied from 70 per cent of the cycle positive to 70 per cent negative. It has been mentioned already that the arc cleaning action takes place only on the electrode positive polarity and that more heat and therefore greater penetration occurs in the work on the electrode negative polarity. Changing the balance between 70 per cent positive and 70 per cent negative gives the option of either maximum arc cleaning or maximum penetration (see Fig. 8.4). For DC use there is often a pulsing facility which allows the welding current to be switched between a low current of say 15 A, sufficient to keep the arc alight and the pulse current of 50–350 A or more. The pulse duration may be 0.05 sec for thin material and up to 5 sec for thick metal. Pulsing techniques are discussed later.

Fig. 8.4 *Unbalancing an AC welding supply to provide either maximum arc cleaning or maximum penetration as inset diagrams show.*

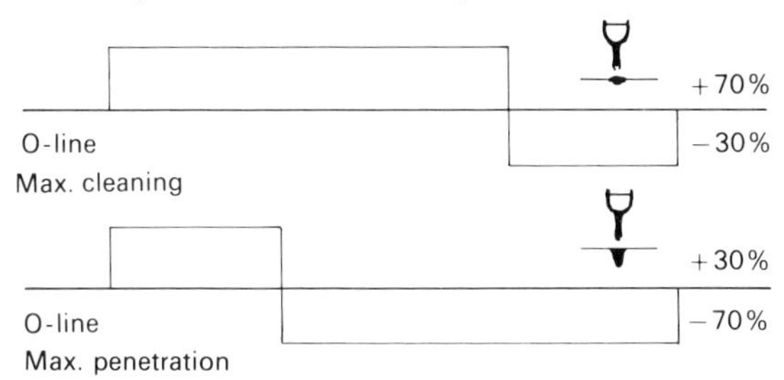

Torches

Heat conducted back up the tungsten electrode must be dissipated otherwise the body of the torch would become excessively hot. Cooling may be by air or water conducted through ducts in the copper body of the torch. Air-cooling is adequate for torches carrying up to about 200 A and above this water-cooling is employed. Frequently the current cable is a bare copper conductor within a hose carrying the water. In this way the size and weight of the conductor can be greatly reduced. Formerly the gas nozzles in TIG welding were often metal and water-cooled but now the nozzles are invariably alumina ceramic. The use of tap-water for cooling the torch body is sometimes inconvenient and subject to freezing in cold latitudes in site work. Self-contained cooler units are therefore used in these circumstances.

Many TIG torches have a switch on the handle of the torch which the operator uses to start the welding sequence. Foot switches are also available. Automatic timing devices are also available for use with special torches designed to make spot welds in overlapping sheets or between sheets and frames when there is no access from the underside. This process variation is called TIG spot welding and requires a torch which is gun shaped and has a nozzle with gas vents which can be pressed on the work. A timed passage of the welding current forms a nugget which penetrates the upper sheet and fuses it on to the underlying member.

To improve the ability to weld in awkward positions some torches allow the angle between the handle and the electrode to be changed quickly (see Fig. 8.5). For welding at the lowest currents a small

Fig. 8.5 *Welding torch with 'bendable' handle.*

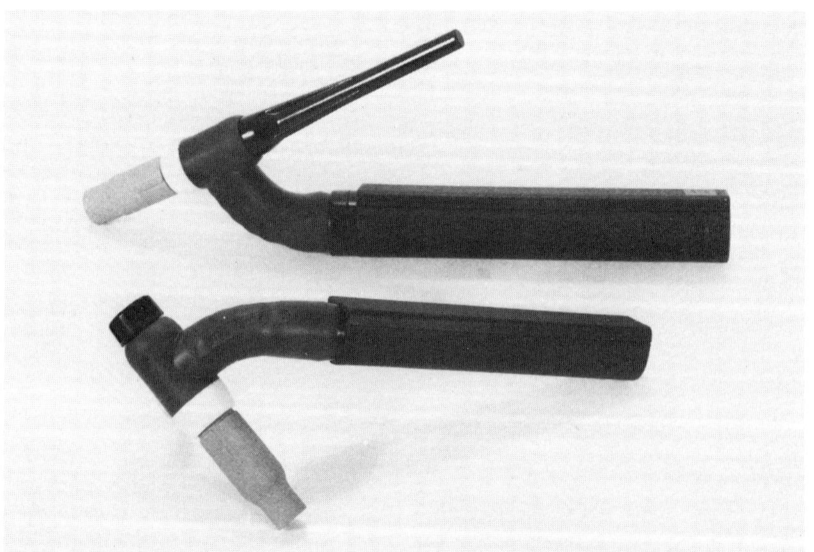

diameter in-line torch, sometimes called a pencil torch, may be used. When access is difficult as in the root of a groove or in a fillet weld it may be necessary to project the electrode well beyond the end of the gas nozzle and this may result in inferior gas shielding because of air entrainment. The difficulty can usually be overcome by employing a gas lens, a special ducted diffuser which fits inside the nozzle and improves the directionality and stability of the gas flow.

Consumables

Gases

Shielding gases are supplied from cylinders or in large workshops from manifold systems. A pressure regulator and a gas-flow meter are necessary at the welding station. The most commonly used gas for TIG welding is argon which can be used on all metals. Argon-hydrogen mixtures containing 2–5 per cent hydrogen are frequently used for stainless steel and nickel-base alloys having the advantage of producing cleaner welds giving deeper penetration because of the higher arc voltage. Hydrogen-containing mixtures must not be used on ferritic steels because of the risk of HAZ cracking or for aluminium and copper alloys because of weld metal porosity. Helium-argon mixtures give deeper penetration, greater heat input and therefore faster welding because of the higher arc voltage than pure argon but arc striking may be more difficult than in argon. These mixtures can be used on alloy and stainless steels and aluminium and copper alloys. Pure helium can be used for welding aluminium and copper alloys, particularly where extra heat input is required in high-speed welding or with heavy plates. Nitrogen, in which tungsten is relatively inert, is not often employed but has been used for copper welding where the high heat input resulting from dissociation of this diatomic gas in the arc and its recombination at the workpiece is useful to offset the high thermal conductivity of the copper.

Filler wires

Generally bare wires of a composition matching the parent metal are used because there are few losses of alloying elements in the arc atmosphere. Additional deoxidants are included in the wires to cope with residual surface oxides or the accidental entrainment of air in the shield. Many of the wires used for MIG and gas welding can also be used for TIG welding. The British Standard Specification for TIG wires is BS 2901: Parts 1–5, the AWS specification for bare wires is A5,2,7,9,10, 14,16,18 and 28.

Applications of TIG welding

TIG welding produces quality welds and is one of the most widely used welding methods both as regards parent materials and the thicknesses which it can tackle although it is less successful with fillet welds. It will weld alloy steels, stainless steels, nickel base and special alloys, aluminium alloys and the reactive metals. With mild and carbon steels care may be needed to avoid porosity and wires with deoxidants are advised. It is usable with a low current on metal down to fractions of a millimetre in thickness but is also used for the root runs in thick metal where its excellent control of fusion allows good shaped underbeads. In its narrow gap variant (see later) it can be used to weld in deep grooves while the hot-wire variant (see later) gives deposition rates competitive with other arc-welding methods. It is frequently used without the addition of filler wire (a technique called autogenous welding, see glossary) especially on thin material, a simplification which is a particular advantage for automatic welding. When filler metal is required it is added to the leading edge of the molten pool and the end of the wire must be kept at all times within the inert gas shield to avoid oxidation.

Arc length must be controlled to close limits which is demanding on the human operator and while it is a process readily mechanised it can only be applied by arc-welding robots which are particularly accurate. In many automatic applications the torch and any filler wire feeding device is held in a stationary fixture with the work moved underneath. Automatic welding heads often incorporate voltage-sensing devices which raise or lower the head to follow the profile of the work and keep a constant arc length.

The TIG process is widely used for the automatic orbital welding of tubes in which a small welding head is revolved around the stationary tube (see Fig. 8.6) and also for making tube/tubeplate joints (see Fig. 8.7 and Fig. 8.8). With orbital welding filler wire may or may not be added depending on the wall thickness of the pipe and whether or not a multipass weld is to be made. In tube/tubeplate welding added filler metal is less common because the joint is usually such that there is sufficient melt-down of either tube or plate.

TIG welding variants

Pulsed current

The use of pulsed current greatly extends the control which can be exercised on the process allowing:

1. Improved consistency in the underbead of unbacked butt welds.

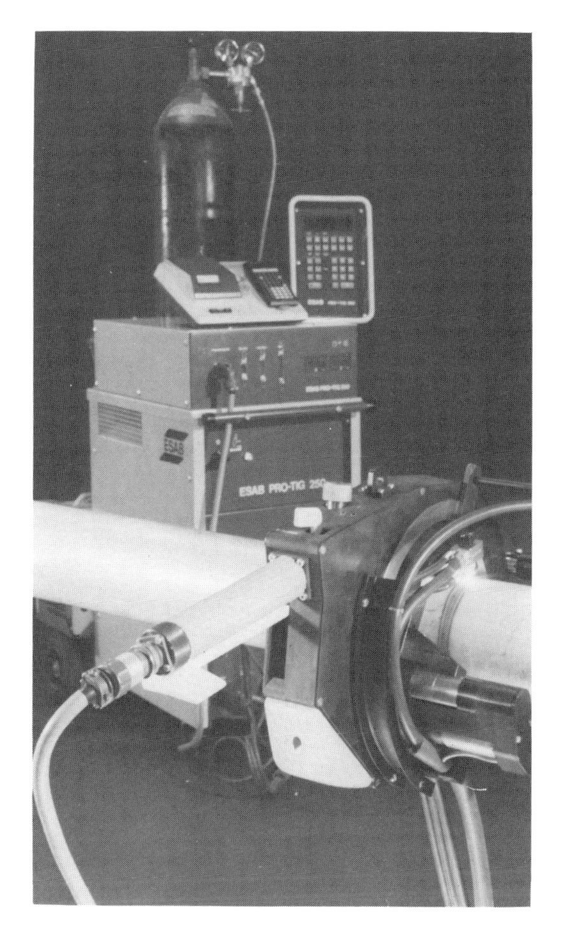

Fig. 8.6 *Orbital TIG welding head with programmable power source.*

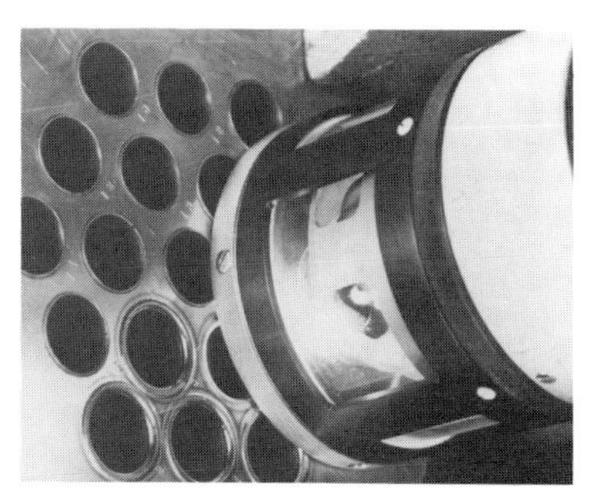

Fig. 8.7 *Two tube plate welders in use on a large tube plate in which a front face weld is made between the slightly protruding tube and the surface of the plate.*

Fig. 8.8 *The tungsten electrode orbits around a locating spigot.*

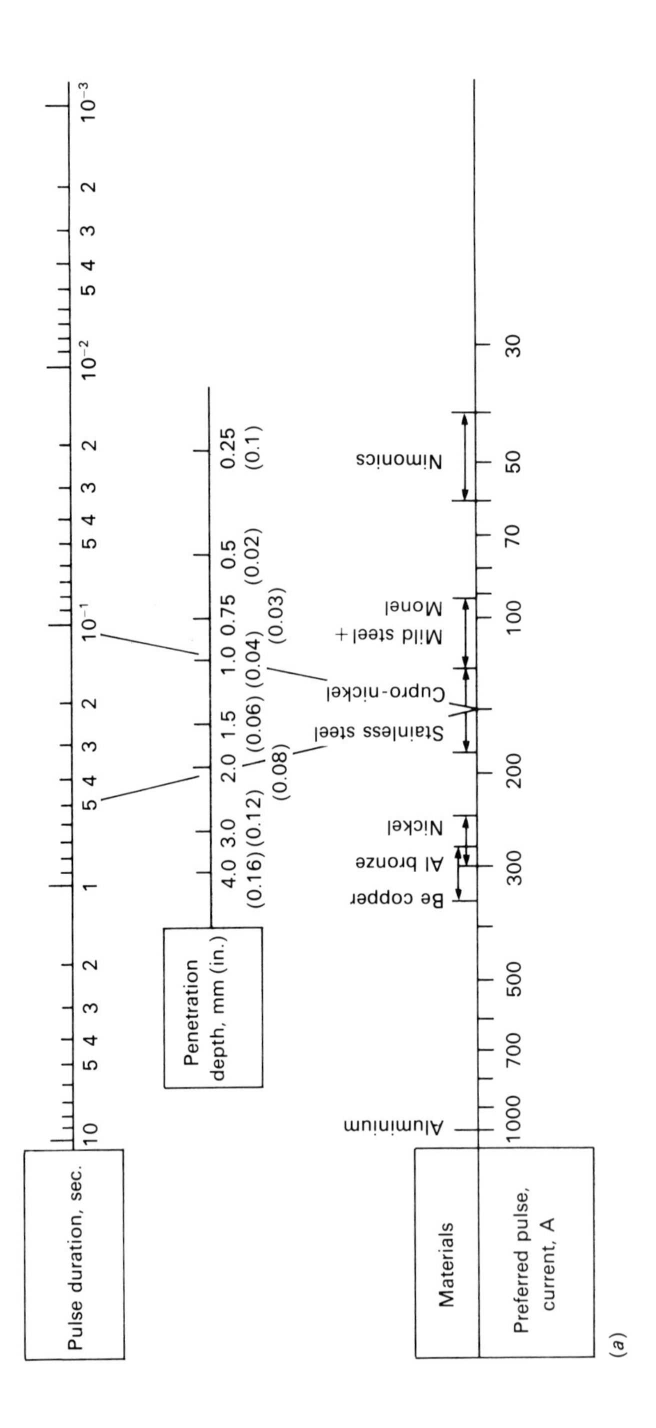

Fig. 8.9 *Nomograph for pulsed TIG welding.*

2. The ability to overcome differences in heat sink and therefore to join thick to thin material.
3. The ability to make cylindrical or circular welds without a build-up of heat and an increase in weld width.

Pulsed arc is able to achieve all these things because the high current during the pulse causes rapid penetration of the workpiece with less spread of heat to the HAZ. Before the pool penetrates the material completely or the weld width begins to increase rapidly the pulse current is cut off and the pool is allowed to solidify. The electrode is then moved on by about half a pool width and the process is repeated to form a series of overlapping spot welds.

The improvement in control results from the compression of the isotherms in the HAZ so that the material close to the weld is not aware of what is happening in the weld!

There are four parameters to pulsing: pulse current, pulse duration, background current and background duration. At first sight this increase in the number of variables is an unwelcome complication but background current is normally around 15 A and the pulse on and off times are usually in the ratio 1:2. Pulse current depends mainly on the materials, those with a high thermal conductivity requiring the highest current and those with low thermal conductivity the lowest. Pulse duration depends on material thickness.

These relationships are summed up in the nomograph in Fig. 8.9. This diagram shows that when pulse welding stainless steel 2 mm (0.08 in.) thick a pulse current of 150 A for 0.5 sec is required. With the same current, 1 mm (0.04 in.) thick stainless only requires a pulse duration of 0.1 sec. Figure 8.10 is an example of pulsed TIG welding. For most

Fig. 8.10 *Pulsed TIG weld between the flexible stainless steel pipe and the mild steel ferrule of a gas fitting.*

applications simple rectangular pulsing is all that is required but for certain demanding applications a shaped pulse having 'slope-out' is found to be necessary.

Hot-wire TIG

The normal way of adding filler metal to a TIG molten pool is to introduce it at the leading edge. If, however, it is fed into the pool itself behind the arc contact can be continuous and it is possible to preheat the filler wire by passing an electric current through it, usually from a separate power source. The arc melts the pool in the workpiece and the pool melts the preheated filler wire. Figure 8.11 illustrates the circuit required, showing an AC power source for the 'hot' wire which is desirable to minimise magnetic interference with the power arc. The extra energy provided by the resistance heating can raise the deposition rate to almost that achievable by MIG welding. This makes it possible to reduce the number of passes in multi-pass welds. The effectiveness of the process depends, however, on the wire between the contact tube and the molten pool having sufficient electrical resistance, a combination of wire diameter and the specific resistance of the wire. Most examples of hot-wire TIG are therefore in steels or high nickel alloys. The process is also used for surfacing operations when twin wires may be used as well as arc oscillation.

Narrow-gap TIG

The reasons for adopting a narrow gap technique have been mentioned in the discussion on narrow-gap submerged arc on page 122. It is estimated that narrow-gap TIG shows an advantage over conventional submerged-arc as regards joint completion time above a thickness of

Fig. 8.11 *Principle of hot-wire TIG welding.*

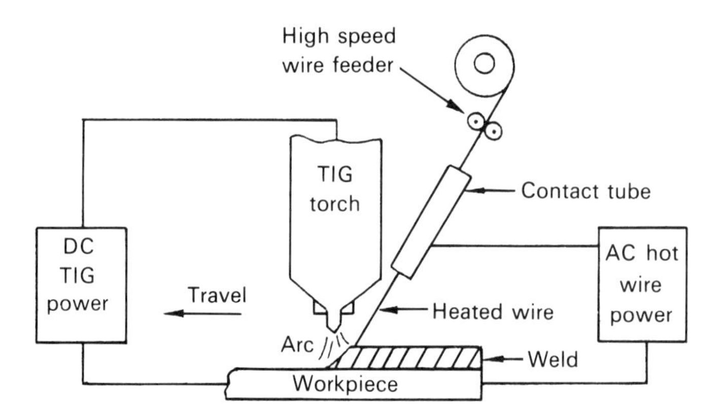

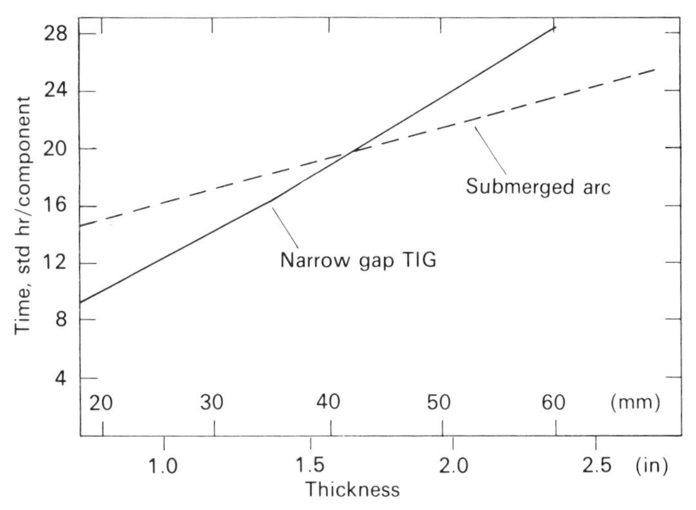

Fig. 8.12 *Relative economics of narrow gap TIG and conventional submerged arc.*

40 mm (1.58 in.), see Fig. 8.12. For thicknesses up to about 20 mm (0.75 in.) a standard torch but with extended electrode stick-out is used but above this thickness a special narrow torch with gas ports fore and aft is required. An arc length control facility is essential. All the special TIG techniques already discussed, such as current pulsing, gas mixtures and hot-wire techniques, have been used for narrow-gap welding. Although the main use has been in the flat welding position on rotated pipe, some users have had successful results with vertical-up welding using the hot-wire technique. The main applications for narrow gap TIG have been cylindrical or circular welds for power generation plant.

Plasma arc welding and cutting

Principles

The plasma arc is a development of the gas tungsten arc and the torches for both processes have similarities as Fig. 8.1 and Fig. 8.2 have shown. With the plasma torch there are two separate gas streams, the plasma gas which flows round the tungsten electrode subsequently forming the core of the arc plasma and the shielding gas which provides protection for the molten pool. The separation of the two flows allows different gases to be used for each function. Argon is the preferred gas for the plasma flow but gas mixtures can be used for the shielding flow although it is also quite common to use argon. Argon-hydrogen mixtures can be used for steels and nickel alloys and helium-argon for titanium and non-ferrous alloys.

The water-cooled copper nozzle and collar through which the arc passes constricts the arc, giving it a columnar shape with a high-temperature core. This arc column is stable and directional; that is, it can be pointed in a given direction even at very low arc currents in contrast with the TIG arc which is broader and not so easily placed on a particular spot. Longer arcs can also be held with less sensitivity to arc gap than with conventional TIG.

Power sources for plasma arcs, especially for plasma cutting, must have higher arc voltages than is necessary for TIG welding, but most modern TIG power sources have a sufficiently high maximum voltage to be used for plasma welding as well. The power sources may be either DCEN or AC square wave, the latter option when employing the unbalanced waveform with 70 per cent positive (see p. 141) makes possible the welding of aluminium alloys. An important distinction from TIG welding is that it is possible to run a pilot arc, indeed the process is started by using a high-frequency voltage to strike a pilot arc between the electrode and the constricting collar. When the plasma gas flow is increased and the torch is brought close to the work the main arc then strikes to the workpiece.

Applications of the plasma arc

The plasma arc is used in four modes:

1. Microplasma welding.
2. Medium current welding.
3. Keyhole welding.
4. Cutting.

Mirco plasma

The plasma arc is stable down to 0.1 A and torches for delicate work have a capacity from this lower limit up to some 20 A. The orifice in the collar is 1 mm (0.040 in.) or less. These small torches giving a needle-like arc are able to weld thin sheet, wire and mesh which a normal TIG arc could not tackle.

Medium-current

Above the micro-plasma range and up to about 100 A the process is 'conduction limited', that is, heat is transmitted from the molten pool to the HAZ by thermal conduction. The collar orifice is now larger in diameter, 1–3 mm (0.040–0.125 in.), depending on the arc current. In this range the process competes with conventional TIG over which it has the advantage of easy arc striking, less chance of tungsten contamination in the weld, a smoother weld finish and narrower, deeper penetration. It finds application in the aerospace and chemical plant industries where it is often used in mechanised or automatic welding plant.

Keyhole plasma

Above about 100 A penetration becomes deep and narrow and it is possible for the plasma to penetrate the workpiece completely. The weld then assumes a wine-glass shape in cross section with heat being introduced throughout the cross-section; it is no longer a conduction limited weld and is decribed as a 'keyhole' weld, Fig. 8.13. (Keyhole welding is a characteristic of electron-beam and laser welding to be discussed in Chapter 10.) The ability to penetrate deeply in this way is partly a result of the concentrated arc giving very high local temperatures and partly a consequence of the kinetic energy imparted by the plasma gas stream. Keyhole welding is possible up to about 6 mm (¼ in.) in many materials but above this limit the hole becomes unstable. The process has been used for welding tubes and other assemblies where access to the reverse side is difficult. When a backing bar is necessary as part of the jigging for work assembly this should not fit closely under the weld as the gas passing through the keyhole (see Fig. 8.14) must have a means of escape. When a weld is terminated by running over the start as in circumferential tube welding a rather complicated procedure is necessary in which both current and plasma gas must be simultaneously reduced. In electron-beam and laser welding the keyhole is formed more effectively in material thicknesses limited only by the power of the equipment and these processes will no doubt provide increasing competition for plasma arc welding.

Plasma cutting

The conditions for keyhole welding are essentially the same as for cutting, the current and gas flow being pushed up so as to cause the

Fig. 8.13 *Comparison of TIG and plasma arc welding. Plasma weld has second 'cosmetic' pass to improve appearance.*

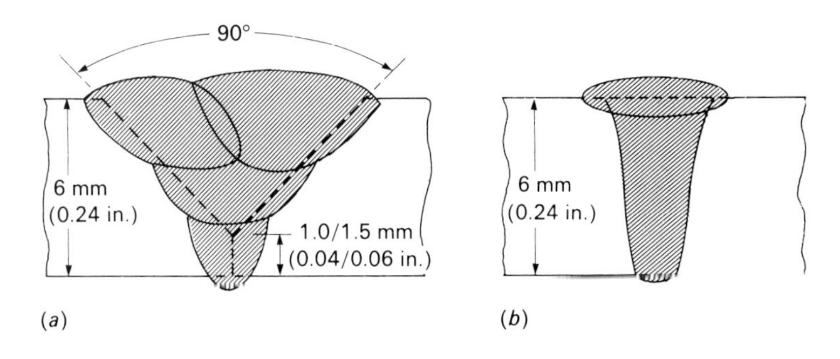

Fig. 8.14 *Keyhole plasma weld in 5 mm (0.20 in.) stainless steel showing the efflux jet. Welding is from right to left side of picture.*

ejection of molten metal from the keyhole. To increase the energy available it is usual to employ an argon-hydrogen gas mixture for the plasma gas. There is a tendency for plasma cutting to produce ozone and oxides of nitrogen but this risk is all but eliminated when the cutting is carried out 75 mm (3 in.) beneath the surface of a water bath. Immersion in water also reduces distortion and the HAZ. Plasma cutting is especially useful for cutting thick non-ferrous metals not readily cut by oxy-fuel gas but it is also used successfully on ferrous metals. As with oxy-fuel gas cutting it is fitted to advanced NC profiling equipment. A variation on the process which is suitable for cutting sheet metal or metal up to 12 mm (nearly ½ in.) thickness at low speed is air-plasma. Although such as aluminium and copper and their alloys. The equipment is essentially similar but with air as the plasma-gas. This necessitates the use of low-plasma currents and special electrodes, of copper with hafnium or zirconium inserts which are resistant to erosion by air. The use of compressed air for both cutting plasma and cooling allows the more complicated and expensive water-cooling to be dispensed with thereby simplifying the torch. The process is operated with the nozzle in contact with the work (see Fig. 8.15).

Fig. 8.15 *Air plasma cutting of a 2 mm (0.080 in.) thickness steel tube.*

Process assessment

TIG and plasma welding are precision welding processes capable of producing high-quality welds in a wide range of ferrous and non-ferrous metals. They exist in a number of forms particularly suitable for use in mechanised and automated equipment. The industries served include: automobile, aircraft engine and aerospace, power generation and chemical plant. Although primarily used on relatively thin materials hot-wire and narrow-gap versions allow surfacing and welding of thick sections normally welded by submerged arc and other methods. Plasma cutting using an inert gas can be used on any metal including non-ferrous metals difficult to cut by oxy-fuel gas methods. Air plasma cutting finds increasing use on sheet steel.

Chapter 9

Gas welding and cutting

In Chapters 4 to 8 the various ways in which an electric arc is used to provide the thermal energy for welding have been considered. There are other forms of thermal energy of which the flame is one of the most versatile. Flames are hotter when the fuel gas is burnt in oxygen than in air because with air each volume of oxygen is diluted by nearly four volumes of nitrogen which takes no part in the combustion but carries away much heat. Of the various fuel gases acetylene produces the hottest flame and is invariably used for gas welding although other fuel gases may be used for lower melting-point non-ferrous metals or for cutting or the allied process of brazing.

The appliance which produces the flame is generally called a 'blowpipe', although the term 'torch' is also used, see note on page 138. Blowpipes are of two types, the original distinction being according to the pressure at which the fuel gas was available. If the acetylene is supplied from a generator in which calcium carbide is reacted with water, formerly a common practice, the blowpipe must be of the injector or low-pressure type in which a high-pressure oxygen jet draws the low-pressure fuel gas into a chamber before both gases pass into the nozzle. Low-pressure blowpipes may also be required with a fuel gas such as town's gas. With gas supplied from cylinders, which is now usual with acetylene, a mixing chamber is inserted between the valves on the

Fig. 9.1 *An oxy-fuel gas blowpipe.*

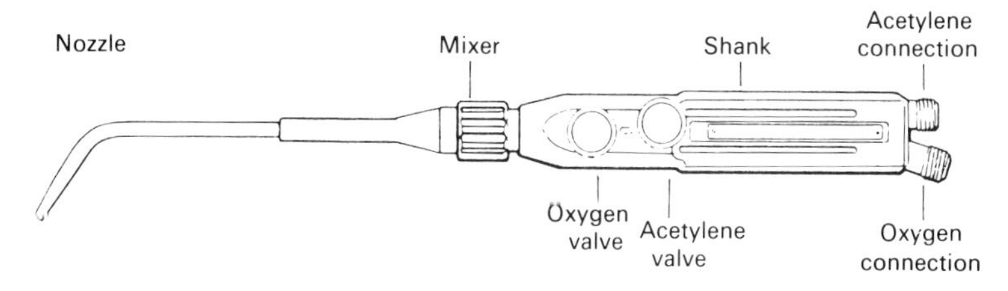

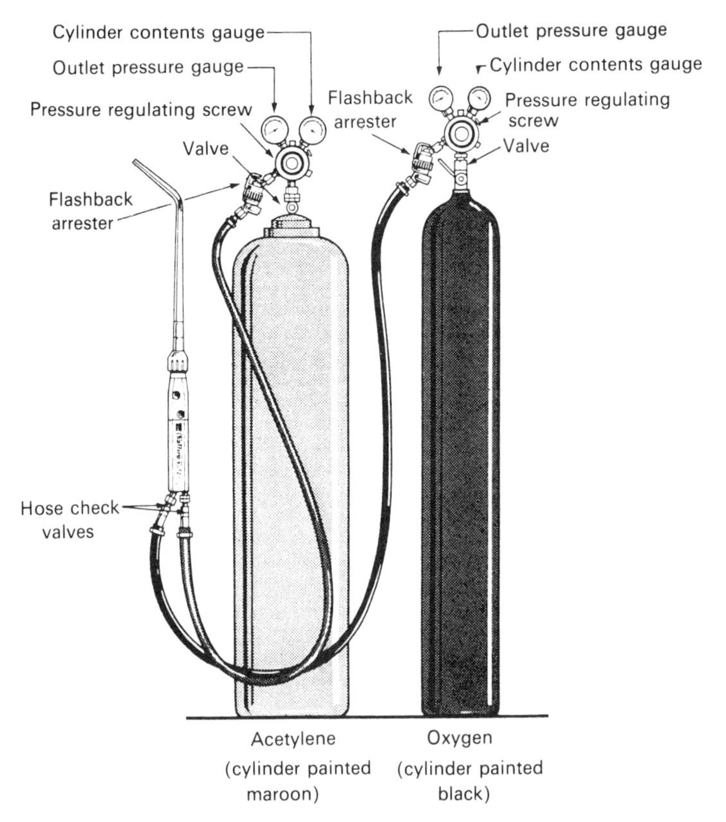

Fig. 9.2 *High-pressure gas welding equipment.*

blowpipe and the nozzle from which the flame is created (see Fig. 9.1). At the present time blowpipes supplied from cylinders may be either injector or high-pressure types, different areas of the world using either depending on the historical development of the process in that area.

The equipment for high-pressure gas welding is illustrated in Fig. 9.2 and includes blowpipe, oxygen and fuel gas hoses, two-stage pressure regulators, the gas cylinders and flashback arresters. Flashback arresters are safety devices which prevent a flame travelling back into the cylinders should a serious backfire occur in the blowpipe. The device extinguishes the flame and makes use of the accompanying pressure wave to operate a mechanism which cuts off the incoming gas supply thereby rendering the apparatus safe. Flashback arresters are a legal requirement in some countries.

Having blown through the hoses with oil-free air (not oxygen) the equipment is assembled and the appropriate adjustments to pressure are made. Equal volumes at equal pressure of the acetylene and oxygen are fed to the blowpipe, the actual pressures depending on the size of the nozzle but being 0.14 bar (2 psi) for the smaller sizes. The flame is ignited after turning on the fuel gas and then the oxygen valve is opened until the desired type of flame is obtained.

The flame has three parts to its structure: a small bright blue-white area near the nozzle called the cone, a bluish area beyond this occupied by the products of primary combustion and an outer envelope which is slightly pink. When there is an excess of acetylene (a carburising flame) free carbon made luminous by the heat causes the central part of the flame to become white and lumnious. As the oxygen valve is opened further this white area shrinks until there is the merest trace of a haze of white on the end of the cone. This is the neutral flame used for most welding. Further addition of oxygen causes the haze to disappear and the cone to shrink slightly. This oxidising flame is used for welding brass or bronze but is not normally employed otherwise. The flame conditions described are illustrated in Fig. 9.3 (a), (b) and (c). The heating power of a flame depends on the rate at which the gases are delivered, that is on nozzle diameter and gas pressure. Gas pressure cannot be varied over a wide range because of its effect on flame characteristics – high pressure a harsh flame, low pressure a soft flame. In practice therefore flame power is proportional to nozzle diameter and nozzles are often described according to the amount of gas they pass per hour.

To extinguish the flame the acetylene is turned off first. If the system is to be shut down for some time it will be necessary to close the cylinder valves and open the blowpipe valves again to relieve the pressure in the system, and also to unwind the pressure screws on the regulators.

When welding mild and carbon-manganese steel the CO_2 and H_2 in the flame are sufficient at welding temperature to break down oxides and produce a clean weld. With other metals, e.g., stainless steel, copper alloys and aluminium alloys, fluxes are required to both clean and protect the weld.

Gas welding is a versatile process capable of welding butt joints without backing and bridging gaps. It is mainly applied manually to sheet material. Where the joint design allows for extra metal to be melted, e.g., in a flanged butt joint, automatic welding may be carried out but this is rare. Automatic operation is frequently used for the allied joining process, gas brazing. The only difference from welding is that there is either pre-placed or added filler metal with a lower melting-point than the parent metal, e.g., copper alloy filler to braze steel. The oxy-fuel gas flame is a diffuse source of heat ideally suited to the brazing process. For welding sheet metal, however, the diffuse nature of the heat source causes a spread of heat and more distortion than with for example the arc welding methods. The blowpipe is widely used in repair

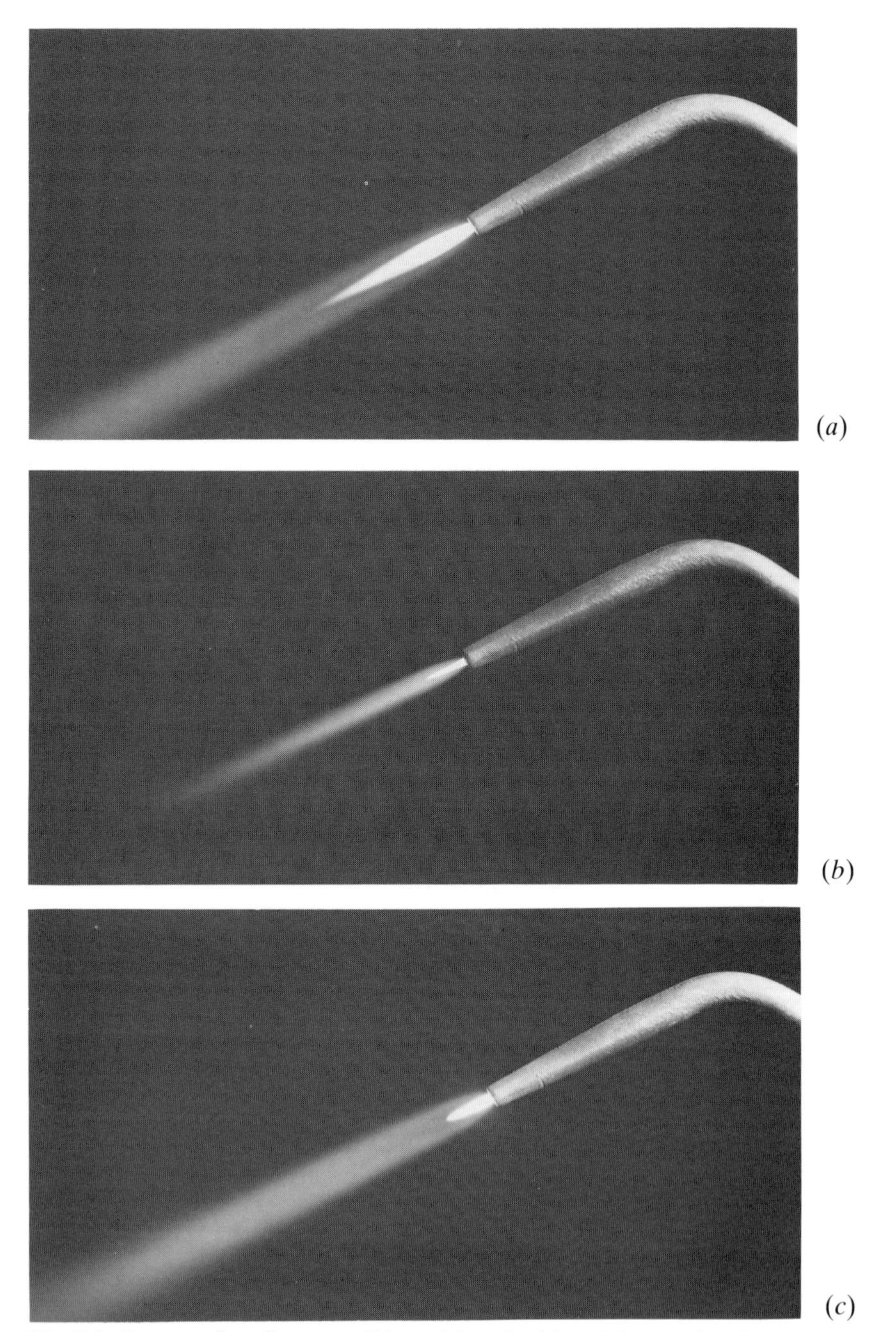

Fig. 9.3 *Oxy-acetylene flame conditions:* (a) *carburising,* (b) *neutral and* (c) *oxidising.*

and maintenance especially for brazing or braze welding. A simple modification allows the blowpipe to be used for cutting.

Special equipment with miniature torches and small cylinders capable of making welds in delicate applications is used in the jewellery trade and for other similar industrial purposes.

Consumables

Gases

Oxygen is available in steel cylinders normally compressed to about 200 bar (2900 psig) or where many outlets are required from liquid stored in refrigerated tanks. Acetylene is made from calcium carbide reacted with water according to the formula:

$$CaC_2 + 2H_2O = Ca(OH)_2 + C_2H_2$$

Acetylene becomes unstable and liable to explode when compressed above 2 bar (30 psig) but can be safely stored under pressure in cylinders packed with porous material soaked in acetone. This is known as dissolved acetylene. Acetylene can be supplied from generators in which calcium carbide is reacted with water. The gas is then at low pressure and an injector type blowpipe must be used. Generated acetylene was once widely used but is rare nowadays.

Cylinders of oxygen and acetylene are often mounted vertically in specially designed trolleys which allows the equipment to be moved close to where it is required. Cylinders must never be used in the horizontal position. Alternatively, a bank of cylinders may be arranged in some convenient position outside the workshop and both gases piped in to distribution points in the workshop. Normally each gas will be supplied from a manifold system with two banks of cylinders, each bank having an isolating valve (see Fig. 9.4). When one bank is exhausted the isolating valve is closed and the one for the second bank is opened. Control systems are available which monitor cylinder pressure and will sound an alarm when the pressure indicates that the cylinder is almost exhausted so that the cylinder banks may be changed over. There are also fully automatic change-over systems.

Filler metals and fluxes

Filler metals are supplied as straight rods in diameters from 1.6–5 mm (1⁄16–3⁄16 in.) and lengths up to 1 m (36 in.). Reels of wire are not used as filler wire additions for gas welding are invariably made manually. Steel wires may be copper-coated to give protection from rusting in storage. For carbon and C-Mn steels the rods usually contain slightly more silicon and manganese than the parent metal and care is taken to ensure

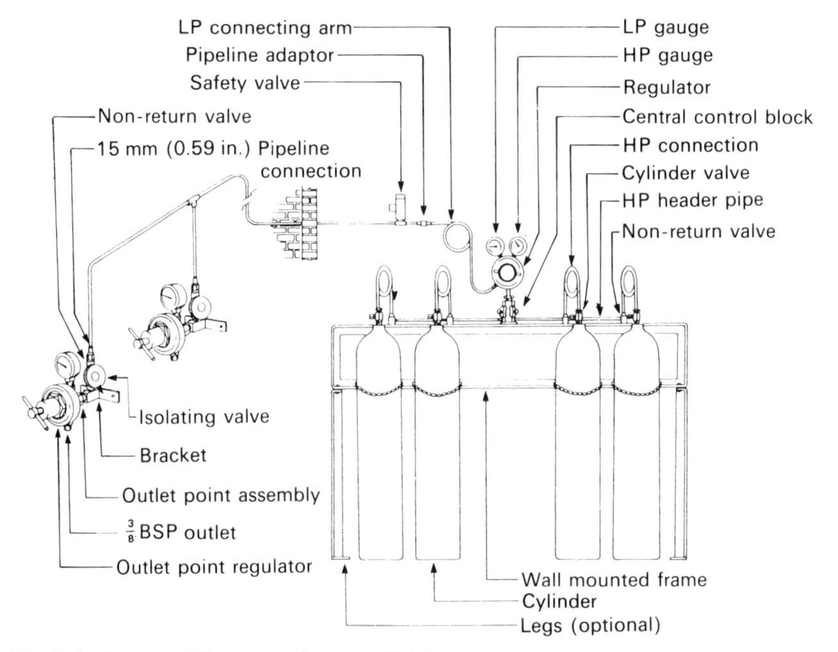

Fig. 9.4 *A gas welding or cutting manifold system.*

that the ratio of Si to Mn in the rod gives a slag which has suitable fluidity. Note that although no flux is used when gas welding mild and carbon steels a very thin layer of oxides form on the surface of the molten pool.

Gas welding is often used for making the root runs in steel pipe because of the excellent control of the unsupported root bead which is possible. For this application filler wires deoxidised with aluminium and titanium or zirconium are sometimes used (the triple deoxidised wire used in MIG welding). Wires containing chromium and molybdenum are available for root run welds in creep resisting pipe steels.

Following the development of the gas-shielded arc processes copper is not welded so frequently by the oxy-acetylene process. The presence of hydrogen in the flame gives a risk of porosity in both weld and HAZ and the high thermal conductivity of copper often makes it difficult to get enough heat into the joint. A copper-silver filler wire (0.5–1.2 per cent Ag) is used for copper welding. Several compositions based on 60/40 brass are used for copper alloys or for repair work, brazing or braze welding (see later). These may contain silicon, manganese or nickel for strengthening purposes. A 10 per cent aluminium bronze, Cu–10 per cent Al, is used in marine work and often in repair work.

The use of gas welding for aluminium alloys has also declined since the introduction of the inert gas arc welding methods, MIG and TIG. It is, however, still used occasionally for pure aluminium with a pure aluminium rod or for the Al–Mn alloy with an Al–Si rod. The Al–10 per cent Si rod melts with fluidity at about 580°C (1075°F), substantially below the melting-point of pure aluminium and the dilute alloys with which it is used for brazing.

The role of gas welding in repair sometimes necessitates the welding of zinc base die castings, for which a matching Zn–4 per cent Al alloy wire is employed. For the same reason rods of either high silicon iron or nickel alloy are available for welding cast iron. Copper-phosphorus alloys containing about 7 per cent with or without silver additions melt with great fluidity at around 700°C (1290°F) and are widely used for the brazing of copper and copper alloys but not ferrous materials, nickel alloys or copper alloys containing nickel.

Standards for filler wires for gas welding and brazing are contained in BS 1453, BS 2901 and BS 1845 and AWS A5.2, A5.9 and A5.10.

No fluxes are required for the gas welding of ferritic materials or when brazing copper with the Cu-P alloy. For aluminium and its alloys fluxes containing chlorides and fluorides of the alkali metals are required to break down the refractory alumina film. These fluxes are highly corrosive and must be removed thoroughly by washing after welding. The design of the joint must be such as to prevent flux entrapment. With other metals fluxes based on boric acid sometimes with additions such as fluorides are used. Fluxes must not only clean and remove oxide films, they are required also to have the appropriate melting-point to allow weld metal or brazing alloy to flow freely. Fluxes can be mixed with water and applied as a paste to the edges to be welded or to the filler rod. Some filler rods are available in which the flux is applied by the manufacturer in the form of a covering to the filler rod rather like a covered stick electrode.

Applying the oxy-fuel gas flame

Welding

There are two major techniques for gas welding, leftward or rightward, which differ mainly in the extent to which the flame points ahead of the completed weld. Leftward welding, Fig. 9.5, is used on steel for flanged edge welds, for square edged plates up to 5 mm (0.2 in.). It is also the method usually adopted for cast iron and non-ferrous metals. Welding is started at the right-hand end of the joint and proceeds towards the left. The blowpipe is given a forward motion with a slight sideways movement to maintain melting of the edges of both plates at the desired rate

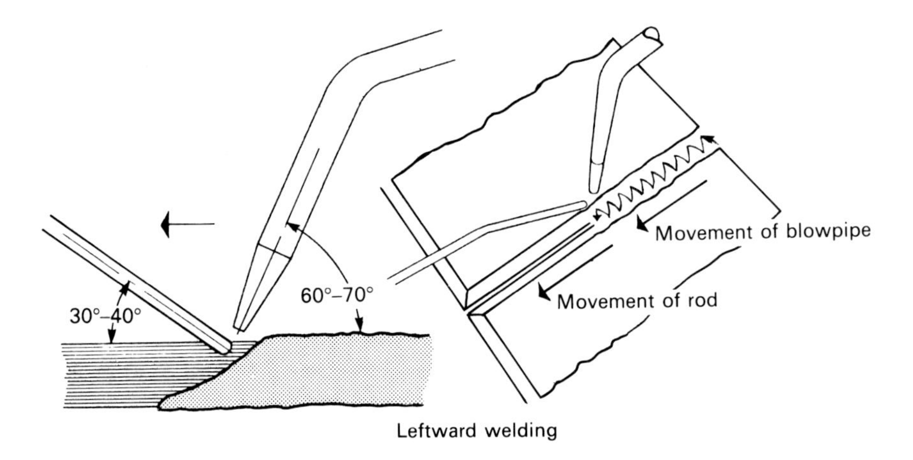

Fig. 9.5 *Leftward welding.*

and the welding rod is moved progressively along the weld seam. The sideways motion of the blowpipe should be restricted to a minimum.

Rightward welding, Fig. 9.6, was the method recommended for steel plate over 5 mm (0.2 in.) thick, but nowadays this thickness would generally be welded by MMA or other processes. Plates from 5 mm–8 mm (0.2–0.32 in.) need not be bevelled; over 8 mm (0.32 in.) the edges are bevelled to 30° to give an included angle of 60° for the welding V. It is suitable for welding in the horizontal/vertical position. The weld is started at the left-hand end and moves towards the right with the blowpipe flame preceding the filler rod in the direction of travel. The rod is given

Fig. 9.6 *Rightward welding.*

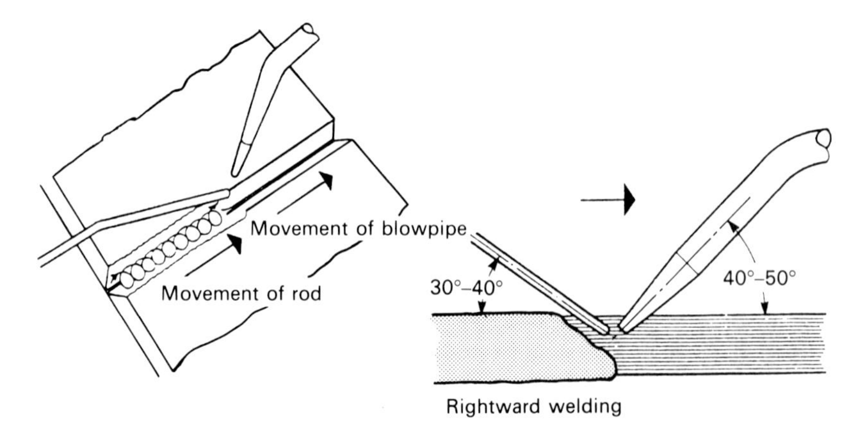

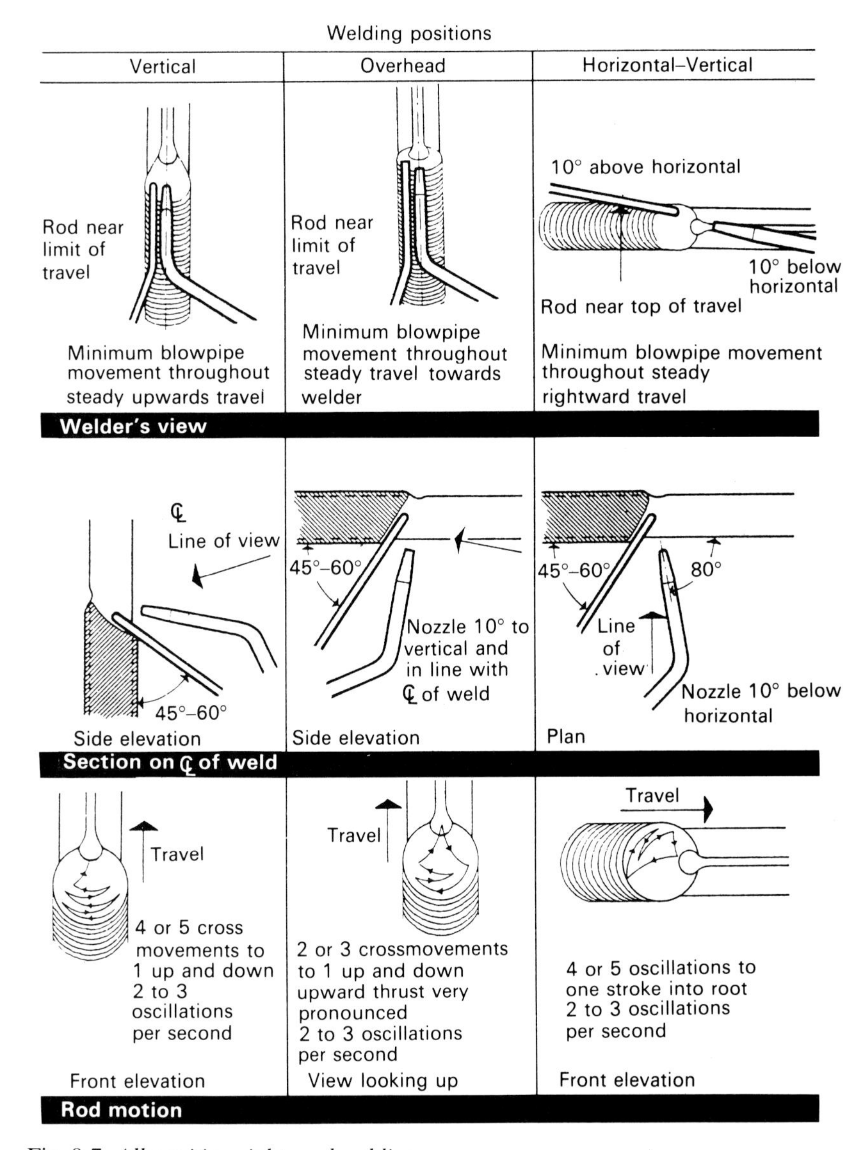

Fig. 9.7 *All-position rightward welding.*

a circular forward motion and the blowpipe is moved steadily along the weld seam. This is faster than leftward welding and consumes less gas; the V angle is smaller, less filler rod is used and there is less distortion.

The all-position rightward technique Fig. 9.7, is a modification of the above and is particularly suitable for mild steel plate and pipe in the vertical and overhead position. The advantages are that it enables the welder to obtain a uniform penetration bead and an even build-up, particularly in fixed position welding and he can work with complete freedom of movement and has a clear view of the molten pool and the fusion zone of the joint. Considerable practice is required to become familiar with this technique even by operators skilled in the normal flat position rightward welding. A form of undercutting of the plate surface at the edges of the weld bead is a fault to which this technique is prone but this can be controlled by appropriate manipulation of the rod and flame. The rod and blowpipe angle should be adjusted to give adequate control of the molten metal as in normal rightward welding.

Vertical welding may be used on bevelled steel plate up to 3 mm (0.12 in.) thickness and has been used up to 13 mm (0.5 in.) by employing two welders working on opposite sides of the joint. When the two-operator method is used, the two welders must be thoroughly trained to work together as a team. With the wide use of MIG and TIG the two-operator gas welding process is rarely practised.

Brazing

In a brazed joint which is always some variety of a lap joint the molten filler metal flows under capillary attraction into small gaps between the surfaces to be joined. The brazing alloy will only flow if the surface is clean, which may require the use of a flux as mentioned in the previous section. Capillary attraction which drives the molten filler metal requires that the gap between the parts is small enough, usually 0.4–0.20 mm (0.008–0.016 in). Heating of the joint should create a thermal gradient up which the filler metal flows into the gap in the overlap. The overlap should be three or four times the thickness of the thinnest member in the joint which should ensure that the joint will not fail in the braze metal. There is normally some superficial alloying between the brazing alloy and the metal in the joint. A joint design which allows the above to occur is vital and joints suitable for welding are not generally satisfactory for brazing, as Fig. 9.8 shows.

Braze welding

This term, also known as bronze welding, is used to describe a joining technique which uses a gas flame and is neither welding or brazing. The same general technique and joint design are used as for welding but filler material is used which melts at a temperature below that of the parent metal. At no time is any of the parent metal melted and in this respect the technique resembles brazing; however, no use is made of capillary attraction (see Fig. 9.8). It is a technique frequently used with great advantage for the repair of steel or cast iron components. The technique is also called bronze welding because the filler metal used is

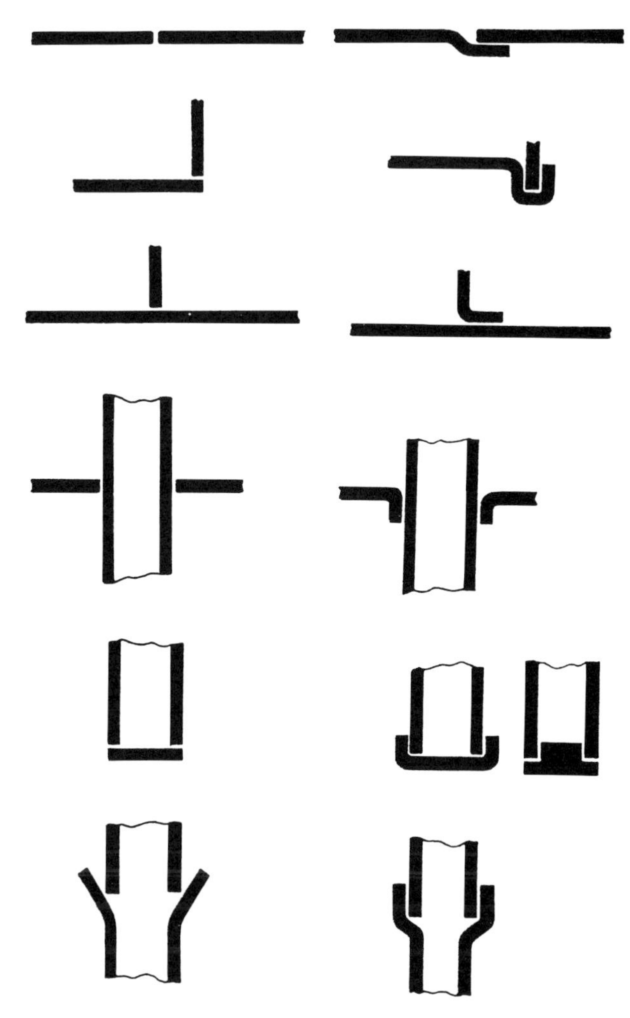

Fig. 9.8 *Joint designs for welding and brazing. (*Left*) for welding and bronze welding. (*Right*) for brazing. (Courtesy of Copper Development Association.)*

generally silicon, manganese, nickel or aluminium bronze. As indicated in the earlier section, of these only aluminium bronze is strictly a 'bronze', the others being 60/40 brasses with alloy additions.

Surfacing

The oxy-acetylene flame is frequently used for surfacing steels and for this purpose this flame has a unique property. When the flame is adjusted to be carburising the excess acetylene reacts with the surface of the steel once it has reached a sufficient temperature. The surface layer becomes enriched in carbon which reduces its melting-point and this superficial fluidity gives the impression that the steel is 'sweating'. After cooling, a steel which has been in the sweating condition has a thin slightly harder surface layer. It is normal, however, to cause the surface to sweat before depositing a thick layer of a hardfacing rod, e.g., one containing carbon and chromium or tubular rods containing tungsten carbide.

Another widely used surfacing method employing the oxy-fuel gas flame is the powder spraying process. While purpose-built spray guns are used the process can be carried out by fitting a suitable attachment to a blowpipe. Metal powder is drawn from a hopper by gas pressure and ejected through a central orifice in the nozzle and thence while molten on to the work surface. A variety of metal powders can be used.

Cutting

The oxy-fuel gas cutting process depends on the reaction between oxygen and iron to form iron oxides. This reaction gives out considerable heat and is self-sustaining once the required preheat or 'kindling'

Fig. 9.9 *Principle of oxy-fuel gas cutting.*

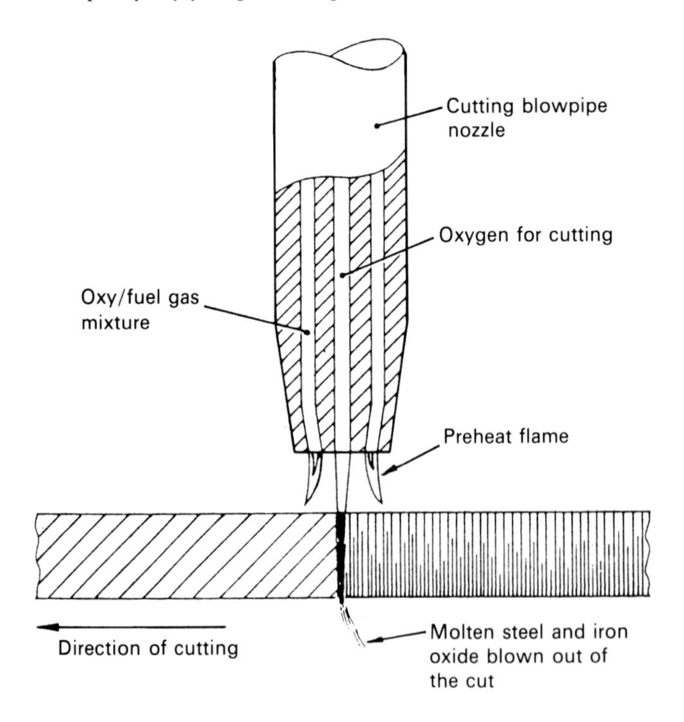

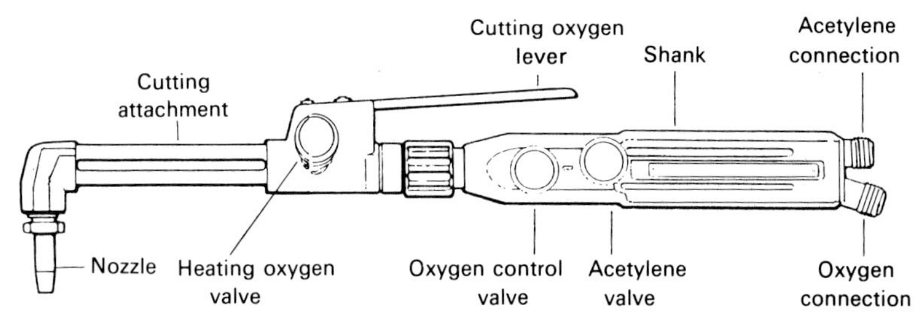

Fig. 9.10 *An oxy-fuel gas cutting blowpipe.*

temperature is reached. To carry out cutting a blowpipe is needed in which the nozzle supplying the oxy-fuel gas flame contains a central orifice through which a jet of oxygen can be directed. Actually the preheating flame is usually supported from a ring of holes as shown in Fig. 9.9. Cutting blowpipes are either combined welding or cutting types or single-purpose cutting blowpipes the latter being usually for heavy duty work. In the combined type the welding nozzle and gas mixer are removed and a cutting head is fitted which includes an additional control valve for heating oxygen and a lever valve for switching on the cutting oxygen (see Fig. 9.10). The heating flame is lit and adjusted to a neutral condition. It is applied at the edge of the material to be cut and when this has been raised to a bright red heat the lever valve is operated allowing the cutting oxygen to flow. Cutting then commences with the flow of oxygen sweeping away the fluid products of combustion through the slot which is being cut. The slot is called a kerf. Oxy-fuel gas cutting is sensitive to the cleanliness of the surface being cut and if possible rust or paint should be removed before cutting begins. Figure 9.11 shows the effects which various changes in operating conditions have on the quality of the cut.

Manual oxy-fuel gas cutting is extremely useful for reclaiming scrap metal and fettling castings, etc. Although manual cutting is used on a small scale for preparing edges for welding the quality of the cut is greatly improved by mechanisation. A major use of oxy-fuel gas cutting is in mechanised or automated cutting. The simplest equipment may have a single blowpipe mounted on a pantograph and moved along a template. At the other extreme a bank of cutting heads with automatic ignition and cutting oxygen control may be operated by a computer-controlled machine with software to allow the nesting of components and a memory to permit standard shapes to be cut on demand (see Fig. 9.12). Gas cutting is often used to prepare plate edges for welding and this requires a cutting head with two or three nozzles to cut the various parts of the preparation.

In the related process of flame gouging the flame does not penetrate

Appearance of cut	Primary cause	Remarks
 Sharp top edge. Smooth surface, drag lines barely visible. A very light scale of oxide easily removed. Square face. Sharp bottom edge.	Correct conditions.	The very light draglines should be almost vertical to profile cutting. For straight cutting a drag of up to 10 per cent would be permissible.
 Melted and rounded top edge. Lower part of the cut face fluted or gouged very irregularly. Bottom edge rough. Slag which is difficult to remove.	Speed too slow.	The bad gouging in the lower half of the cut is caused by molten steel scouring the cut surface and the hot metal and slag which congeals on the underside is always difficult to remove. Secondary cause of this condition is oxygen pressure being too low.
 Top edge not so sharp and may be beaded. Undercutting at top of cut face. Drag lines have excessive backward drag. Slightly rounded bottom edge.	Speed too fast.	The excessive backward drag of the cut line would result in the cut not being completely severed at the end. The occasional gouging or fluting along the cut indicates that the oxygen pressure is too low for the speed, but possibly not too low for a normal speed. In other words, if the speed was dropped and the oxygen pressure maintained, a perfectly good cut would result.
 Rough surface, poor appearance, top of face like folded curtain with beads on lower edge. Flat face has vertical runs. Slag adheres to underside.	Dirty nozzle.	Clean cutting orifice carefully with correct cleaning tool.
 Rounded top edge with melted metal falling into kerf. Cut face generally smooth, but tapered from top to bottom. Excessive tightly adhering slag.	Preheat flame too large.	This is the easiest and most obvious condition to correct. Providing other conditions are normal the appearance is of a clean but heavily oxidised face combined with very heavy rounding at the top edge.
 Regular bead along top edge. Kerf wider at top edge with undercutting of face just below	Pressure of cutting oxygen too high.	Probably the commonest fault in cutting, causing rounding of the top part of the cut face through turbulence within the oxygen stream which is set at too high a pressure. On thinner material it may cause a taper cut which sometimes leads to the incorrect supposition that the cutter is incorrectly mounted in relation to the plate.

Fig. 9.11 *Cutting conditons and cut quality.*

the work but with the aid of a special nozzle scoops metal from the surface. The technique is used for removing defects from the surface of ingots and for the back gouging of welds, etc.

Fig. 9.12 *A computer-controlled multi-headed cutting machine.*

Process assessment

Gas welding is generally used manually. The equipment is simple, inexpensive, reasonably portable and can be used where electric power is not available. Excellent control can be obtained over the molten pool so the process is often used for making unsupported butt welds and the root runs of pipes. A variety of metals and alloys can be welded though the list is not so extensive as with TIG welding. Special miniature apparatus is used for delicate welding and brazing operations. The oxy-fuel gas heat source is also used for brazing, which is frequently mechanised using pre-placed filler metal. The oxy-fuel gas flame is also used for surfacing. Because of the diffuse nature of the heat source and the inefficiency of heat transfer compared with an arc the welding of thick material tends to be slow and distortion is often serious. The flux which is necessary with some non-ferrous metals such as aluminium is a corrosion risk.

The most important use of the oxy-fuel gas flame is for preheating in the oxygen cutting process. This is a widely used process for preparing edges for welding and shaping section, tube and plate parts for welding. Large highly developed computerised cutting plants are in use in the shipbuilding and structural engineering industries.

The oxy-fuel gas flame is an indispensable source of heat for a variety of non-welding jobs in the workshop from preheating for welding and straightening to flame cleaning.

Chapter 10

Welding and cutting with power beams

Both electron and laser beams have become increasingly important as energy sources for welding and cutting. These two energy sources have many features in common, the most obvious being that the energy is directed to the workpiece as a focused invisible beam and that heat is only released where the beam strikes the work. This feature has resulted in the heat sources becoming known collectively as Power Beams.

When the power density at the focus with both energy sources exceeds about 10 kW/mm^2 (6.452 MW/in^2) energy is arriving faster than it is conducted away as heat. As a result vaporisation takes place forming a cavity in the workpiece which allows 'keyhole' welding to take place. Figure 10.1 illustrates the principle of keyhole welding in which the

Fig. 10.1 *Principle of keyhole welding:* (a) *close butt joint,* (b) *power beam,* (c) *molten metal,* (d) *energy passing through hole,* (e) *full penetration weld,* (f) *welding direction,* (g) *solid weld bead, and* (h) *keyhole.*

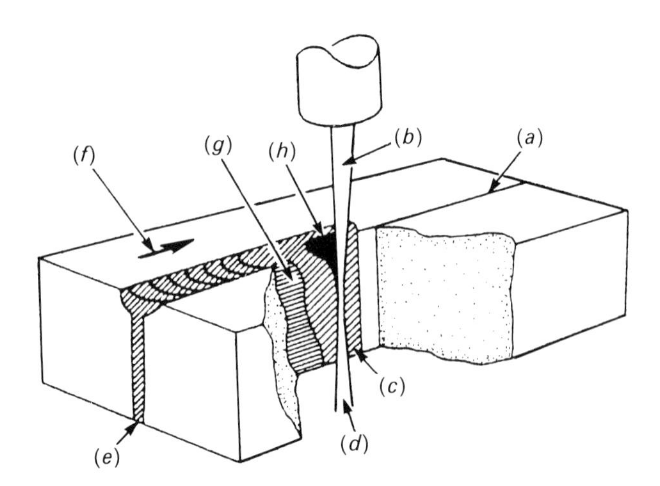

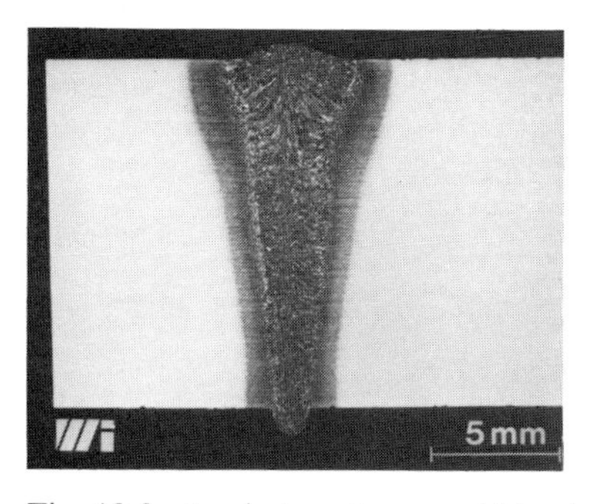

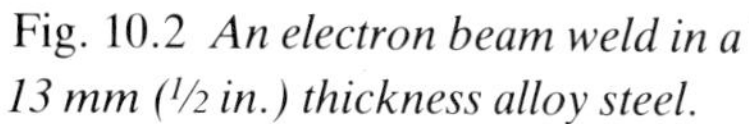

Fig. 10.2 *An electron beam weld in a 13 mm (½ in.) thickness alloy steel.*

Fig. 10.3 *A laser weld in a 10 mm (0.4 in.) thickness alloy steel.*

cavity is traversed through the workpiece, the molten metal lining the cavity flowing behind to form the weld. Figure 10.2 and Fig. 10.3 show electron beam and laser welds made by keyholing. The higher the power in the beam the faster the welding can be carried out. Although at low total powers of about 1–2 kW the penetration and welding capability of the two energy sources is comparable, as the power is raised to cope with thicker material the electron beam becomes increasingly more efficient than the laser.

At the present time lasers are readily available with powers of up to 10–15 kW and electron beam welders up to 100 kW. The electron beam therefore has the potential for welding massive thicknesses in general and power generation engineering. It is also at 75 per cent an efficient converter of electrical energy into heat. In contrast the laser is extremely effective up to a thickness of 12 mm (nearly ½ in.) but it is a relatively inefficient converter of electrical power yielding only 10 per cent overall. With both energy sources, however, the extremely high energy concentrations in the workpiece allow high welding speeds and maximum possible thermal efficiency to the melting of the molten pool. In practical terms this means that the melting of the pool is accomplished with the minimum loss of heat into the workpiece as a heat affected zone. This, combined with the deep, narrow, parallel-sided welds produced by keyholing allows welds to be made with negligible distortion. It is possible in fact to use the processes to join together finished machined parts such as gear clusters.

Welding with both heat sources is fully mechanised, the equipment having a higher degree of precision than other types of welding plant because of the high speed required and the close tolerances necessary to use the narrowly focused power beams. Both heat sources can be used for hole boring and surface treatment but these functions are now invariably carried out by the laser, mainly because it operates in air whereas the electron beam must be used inside a vacuum chamber. The laser also has an important role as a precision thermal cutting device for straight and profiled cuts in both metals and non-metals. Both processes are clean precision techniques for metal working and the assembly of engineering components.

Electron-beam welding

Equipment

An electron-beam welder, even one of 100 kW capacity has the same general principles as a video tube operating in the milliwatt range, that is, an electron gun, focusing coils, deflection coils and a vacuum enclosure. The electron gun resembles a thermionic valve and has at one end a heated filament, ribbon or some other source of electrons surrounded by a bias shield and at the other end an anode with a central orifice. A positive voltage from 30,000–200,000 V accelerates the electrons through the anode to be focused by a magnetic lens. Beyond the lens are two sets of deflection coils by means of which the beam may be positioned or caused to execute a pattern or raster movement. The electron beam must be generated in a vacuum for two reasons (a) protection is required for the hot cathode, and (b) the electron beam would be dispersed by collisions between the electrons and molecules of air. A hard vacuum (10^4 torr) is required in the electron gun itself but a soft vacuum (10^2 torr) can be accepted in the work chamber. It is therefore usual to have separate vacuum systems for gun and chamber. The essential features of the system are shown in Fig. 10.4. Except for work on a micro scale the deflection coils are not used to move the beam over the work. Generally the electron gun is stationary and the work is moved on NC manipulators within the vacuum chamber. With the gun

Fig. 10.4 *Principle of electron-beam welding.*

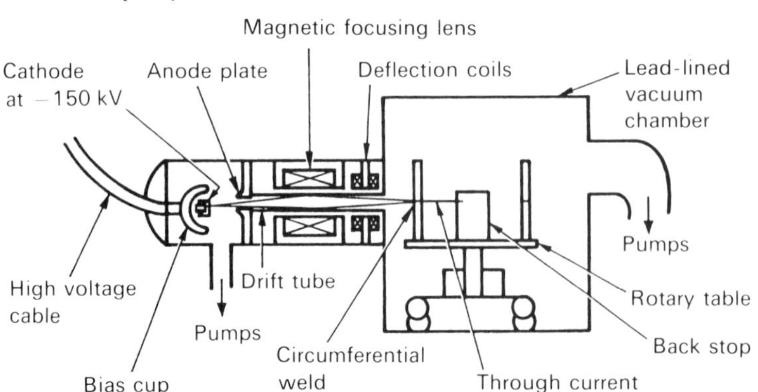

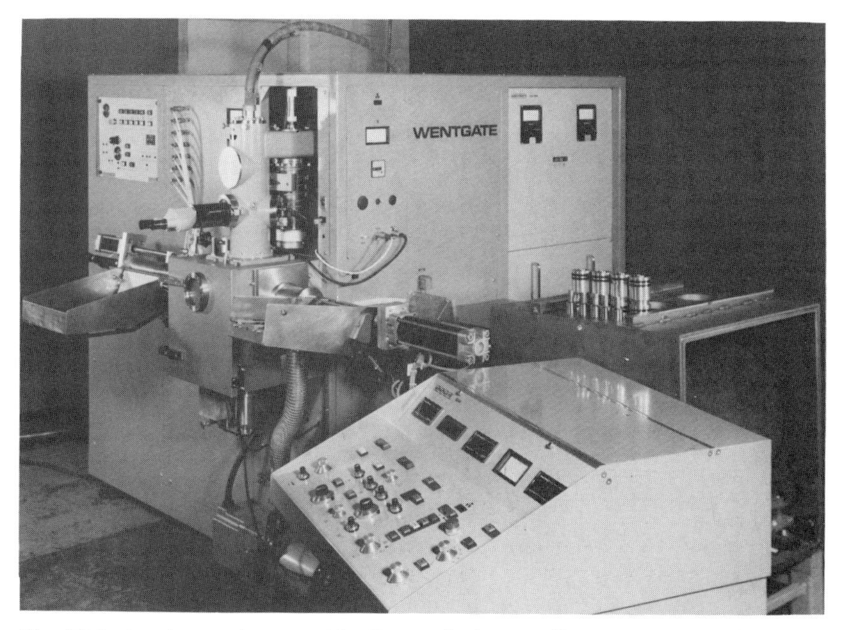

Fig. 10.5 *An electron beam welder for producing small components continuously. The parts to be welded are assembled in jigs in cylindrical capsules with 'O' ring seals at each end* (centre right)*. These are pushed through a tube* (centre) *into the vacuum of the working area where the component is welded. The capsule is then pushed out of the exit tube* (left) *and the finished component is removed.*

stationary and mounted outside the chamber it is necessary to have a long working distance but fortunately with a suitable gun design electron beams can be thrown for distances of up to about 1 m (40 in.).

There are no other welding processes in which such a wide range of power from less than 1 kW to 10s of kW can be delivered by a single equipment. Electron guns and chambers tend, however, to be designed for a particular range of operations. For welding large numbers of small components a large batch is often loaded into the chamber and the components brought one at a time under the welding station. With larger components a sliding-seal device may be used to load components into and out of the welding station. Heavy parts or those on which there is to be much welding are welded one at a time. Chambers may be designed to accommodate particular shapes in such a way that the space to be evacuated is minimised. Figure 10.5 and Fig. 10.6 are of industrial electron beam welders.

In a special adaptation of the process called 'non-vacuum' electron beam the beam is brought out into the atmosphere through a series of orifices and finally screened in a jet of helium. The beam is rapidly

Fig. 10.6 *A 150 kV 8 kW electron beam welder with a 1.7 × 1.2 × 1.0 m chamber. It has full computer control of beam parameters and work handling. A sliding door opens and there are rails (*right*) which are designed to receive workpiece and tooling when they are run out of the chamber. (A smaller machine is in the background.) (Courtesy of Wentgate Dynaweld.)*

degraded by collisions with air molecules and the work must be close to the final orifice. However, it is still an intense heat source and has found application in the continuous welding of sheet.

When the electron beam strikes the work X-rays are emitted which are screened by the steel of the vacuum chamber. With non-vacuum welding the X-rays must be screened off by placing walls round the working area. Even with in-vacuum welding it may be necessary with high voltage guns to take special care in designing the chamber to prevent leakage of radiation.

To increase the volume of the chamber which can be covered by the electron gun it may be brought inside the chamber and provided with various degrees of motion. This means having a flexible high tension cable within the chamber necessitating as a rule a lower voltage (30–60 kV) gun with the penalty of a shorter focus distance. Systems for controlling motion are as varied as those used on machine tools. Increasingly, computer control is employed because of its adaptability and the ease with which simultaneous motion on more than one axis may be controlled. Computer control is also necessary when a seam-following device is employed. One type of seam follower uses back-scattered electrons to provide an image of the weld area.

Welding techniques

Electron-beam welding is ideal for making square close butt joints in a single pass and thickness exceeding 150 mm (6 in.) of steel can be welded with adequate power. Up to about 50 mm (2 in.) welding is done in the flat position but on material much thicker than this surface tension will not hold up the column of liquid metal in the keyhole making vertical or HV position welding necessary. An idea of the capability of the electron beam to weld different thicknesses can be obtained from Fig. 10.16 on page 182, which shows that the process is up to ten times faster than any other arc method. Joints must fit closely and they are usually machined. It is possible to weld with a small gap but as filler metal is not normally added the result will be that the top of the bead sinks. With material of modest thickness the welds can be made to have an excellent top bead especially if a final smoothing or 'cosmetic' pass is made but at a greater thicknesses there is a tendency for spatter to appear on both top and under surfaces. The technique adopted to overcome this is to provide on the underside either integral or separate backing pieces while on the topside an integral flange is incorporated. When welding is completed both top flange and backing are machined away. The provision of a backing as shown in Fig. 10.7 also enables the parts to be located accurately for welding.

Fillet joints require care in design as no metal will be added to create a fillet. The best approach is to turn the joint into one that can be made

Fig. 10.7 *Joint design for an electron beam weld incorporating both flange to provide extra metal and an integral backing bar.*

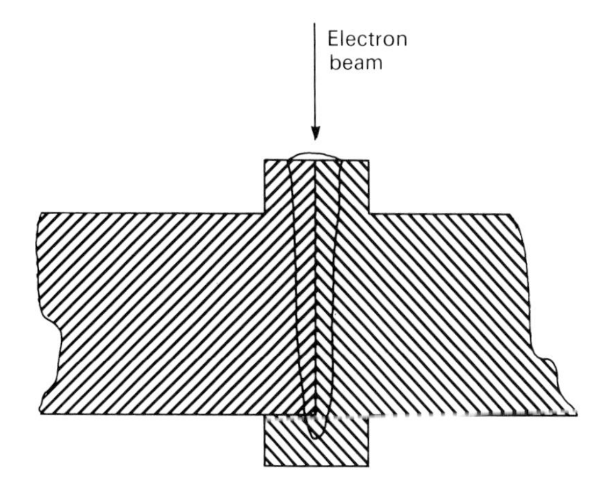

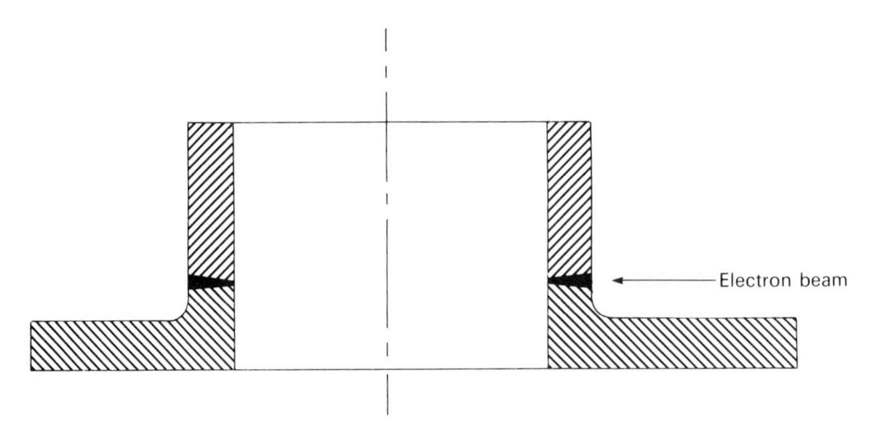

Fig. 10.8 *Joint design for flange or T joints to be made by electron beam. The design shown gives the highest integrity joint.*

by one or more butt welds. Figure 10.8 illustrates a suitable flange-to-tube joint.

It is possible to make welds which are extremely narrow but this creates the risk of several kinds of weld defect. Root porosity and ragged penetration are particularly likely and lack of root fusion and cavities further up the weld also occur. If the weld is made broader and less pointed and beam spinning is applied porosity is able to escape and sound welds can be made. The spinning or oscillation of the beam is carried out at about 500 Hz. Beam spinning is also desirable together with power reduction at the overlap point in weld completion with a circumferential weld.

The rapid and local heating and cooling experienced by an electron-beam weld results in harder weld metal and HAZs in alloy and hardenable steels than are found in arc welds. This may require the completed parts to be heat treated after welding or if the parts are small the use of the electron beam itself as a heat source. The latter course of action may well be uneconomic, however, because it takes up time in a costly item of plant.

Weld metal or solidification cracking which in arc welds is strongly influenced by the consumables is totally dependent on the analysis of the parent metal. Attempts have been made to place shims of an appropriate composition in the joint to provide control of composition but the best solution is to ensure that the parent metal does not have a crack-sensitive composition. In certain highly alloyed materials it may be that limiting the welding speed will effect an improvement. Electron beams are deflected by magnetic fields and this can cause a serious lack-of-fusion defect in thick welds in which the top of the joint is correctly fused but the root is missed entirely by the weld. It is necessary when plates contain residual magnetism to degauss them thoroughly before welding. Similar problems arise when dissimilar metals are welded because thermo-electric currents are generated between the two parts of the weld. There is no pretreatment in this case which will prevent the beam deflection.

Applications

Although initially it was used mainly for nuclear engineering and the aerospace and aircraft engine industries, where quality and lack of distortion were of paramount importance, the process subsequently found growing use in the automobile industry. Here it justified its use in engine and transmission parts on the basis of welding speed and overall economics (see Fig. 10.9). It is capital-intensive equipment but the ability to make quality welds with low total heat input and distortion as well as to join parts of dissimilar thickness allowed designers to redesign parts with fewer welds or even to fabricate in novel ways. The link with the designer must be stressed as there are usually few advantages in merely replacing arc welding by electron beam. With the advent of beam powers capable of welding massive thicknesses the process is becoming used in the manufacture of power generation heavy electrical and chemical plant.

Fig. 10.9 *Components for an automatic gearbox made by electron beam. (Courtesy of Wentgate Dynaweld.)*

Laser welding and cutting

Types of laser

There are several types of laser but only the solid-state YAG (Yttrium-Aluminium-Garnet) and the carbon dioxide laser are of significance in metal or materials working. These lasers take their names from the lasing medium in the optical path between two mirrors, the one on the output side being only partly reflecting. This lasing medium is caused to emit stimulated radiation the axial component of which is reflected to and fro between the two mirrors being amplified in the process. The laser beam is emitted through the part silvered mirror and is light of a single wavelength with all waves in phase, that is, it is a coherent light beam. The process necessary to stimulate the emission in the lasing medium is: for the YAG laser the light from one or more powerful flash tubes, and for the CO_2 laser an electrical discharge of some tens of thousands of volts.

The solid state YAG laser

This laser emits in the near infrared region of 1.06 μm, a wavelength readily absorbed by metals and capable of being focused by ordinary lenses. Normally the input and output are pulsed and can deliver up to almost 500 W. The heat generated by the flash tube in the cavity is usually removed by water cooling and the whole equipment has an efficiency of only 4 per cent.

Solid-state laser equipment is illustrated diagrammatically in Fig. 10.10 and an actual equipment is shown in Fig. 10.11. Such equipment would have an output with pulse lengths of 1.6 milli sec with a 1 sec frequency and give a spot weld of 0.2–1.0 mm (0.008–0.040 in) diameter. While it is possible to make continuous cuts and welds by an overlapping technique the solid-state laser is more normally employed in operations such as hole drilling, resistor trimming and the welding of small components and micro-electronic circuitry. Hole drilling is a particularly attractive application as the process is fast, can work down to very small diameters and is not sensitive to the hardness of the workpiece.

The carbon dioxide laser

The output of this laser is in the mid infrared, 10.6 μm wavelength requiring optics in potassium chloride or zinc selenide. It is a wavelength less readily absorbed by metals than the output of the solid-state laser until the melting point is reached. CO_2 lasers are available with powers from a few hundred watts up to 20 kW. Although it can be operated in a pulsed mode it is normally a continuous output which is particularly suitable for welding and profile cutting. The lasing medium is gaseous, a

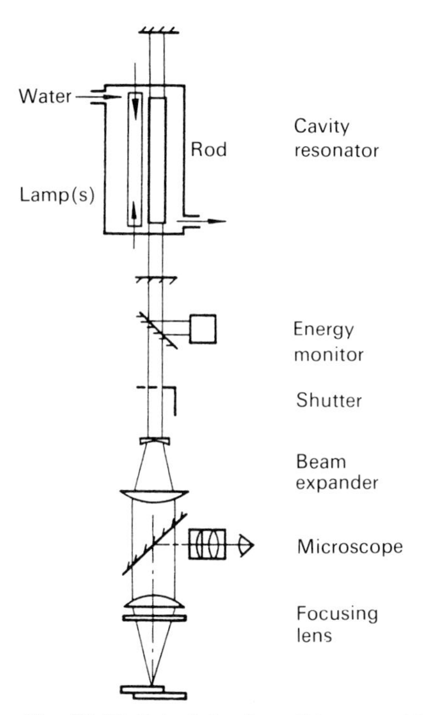

Fig. 10.10 *Principle of a solid state welder.*

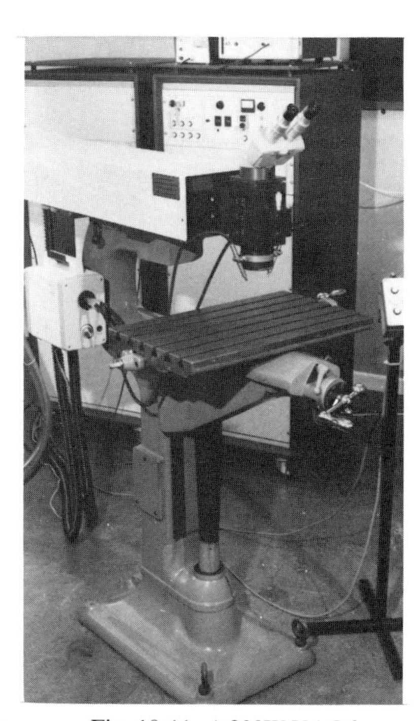

Fig. 10.11 *A 300W YAG laser.*

mixture of carbon dioxide, nitrogen and helium at a reduced pressure of 20–50 torr.

Several forms of CO_2 laser exist according to how the problem of removing unwanted heat from the laser cavity is tackled. Up to an output of 100 watts the unit may be completely sealed. Above this figure and up to about 500 watts slow flow systems may be employed in which the lasing gas flows through the glass tube cavity, often to waste, while the heat is removed from the water-cooled tube walls. A fast axial flow of gas permits powers up to about 6 kW. Gas is circulated by a Rootes blower at near supersonic speed through the cavity and a heat exchanger. In the transverse flow laser the gas is circulated across and not through the length of the cavity as with the slow flow and fast axial flow systems. The gas is therefore in the cavity for a shorter period of time and is circulated by a fan. Powers of up to 25 kW can be obtained from transverse lasers. The principles of fast axial flow and transverse flow types of laser are illustrated in Fig. 10.12 and Fig. 10.13, and an actual 2.5 kW fast axial flow laser in Fig. 10.14.

Applications of the laser

As an intense, clean and easily directed heat source the laser finds use in

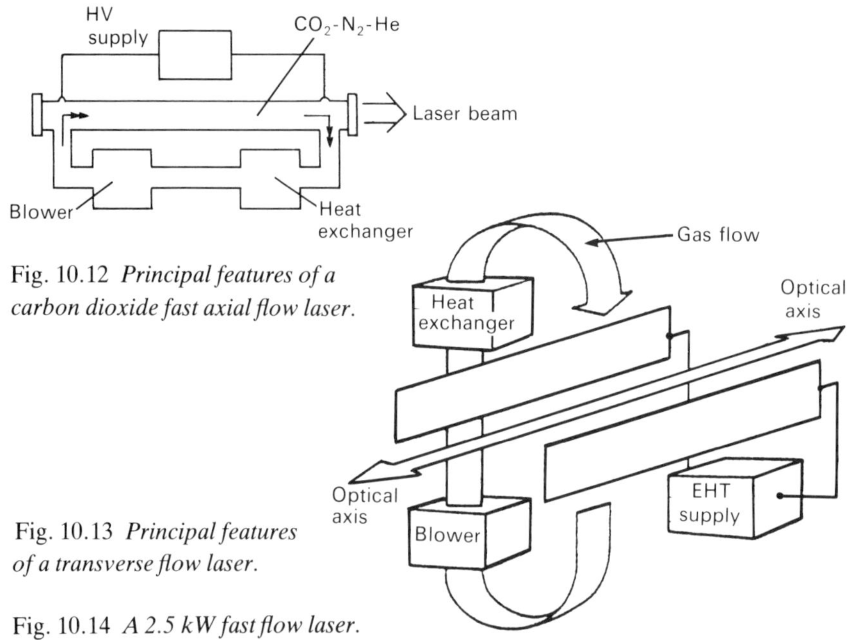

Fig. 10.12 *Principal features of a carbon dioxide fast axial flow laser.*

Fig. 10.13 *Principal features of a transverse flow laser.*

Fig. 10.14 *A 2.5 kW fast flow laser.*

a variety of metal working applications such as, cutting, welding, surface marking, heat treatment and surface alloying.

Laser cutting

Laser cutting was the first and still is the most important application for CO_2 lasers up to 500 W. It is normally carried out with the aid of a gas jet, as shown in Fig. 10.15, oxygen or compressed air being used for ferrous materials (as the resultant exothermic reaction assists the cutting process) and inert gases when cutting non-metals. The process will cut profiles with precision at high speed and is especially suitable for heat resisting and alloy steels which are difficult to cut by other means. Cutting speeds for mild steel with a 500 W laser are indicated by the rule-of-thumb:

Steel thickness	mm	1	2	3	4	$\frac{in}{1,000}$	40	80	120	160
Cutting speed	m/min	4	3	2	1	ipm	160	120	80	40

Laser welding

Up to nearly 1 kW power laser welds tend to be conduction limited but

Fig. 10.15 *Principle of laser cutting.*

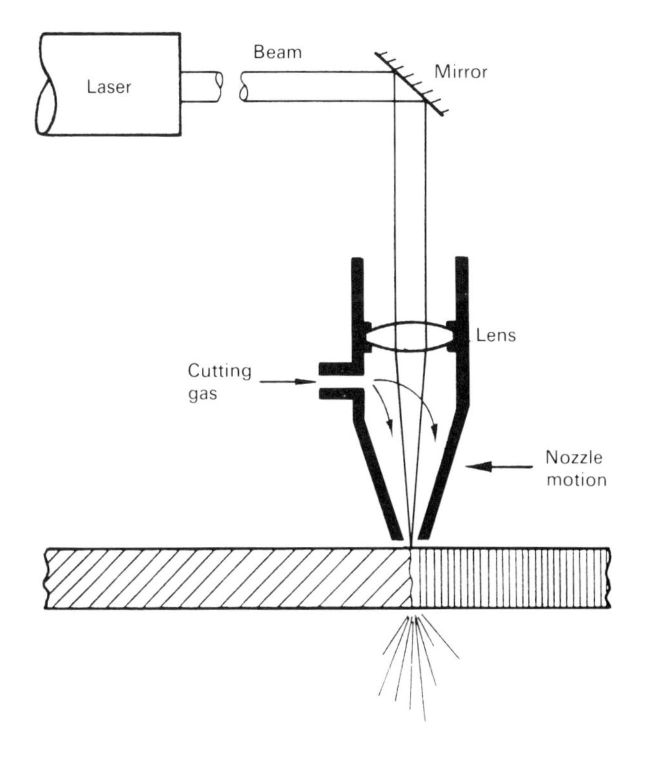

above this power keyhole welding takes place and deep narrow welds similar to those produced by electron beam are possible. The comparison with electron beam is made in Fig. 10.16 and the main reason for the poorer performance of the laser, power for power, is that metal vapour in the form of plasma above the cavity interferes with the laser beam. One way of minimising the effect of the plasma is to ensure that welding is carried out at the highest possible speed as in this way the plasma cloud is left behind. Other techniques for avoiding the effect rely on blowing the plasma away with a gas stream.

Laser welding with powers up to 5 kW is being successful in the same general areas as low-power electron beam, e.g., in aerospace and automotive manufacture. The laser is often preferred to electron beam because welding can be carried out without a vacuum chamber. This reduces considerably the complexities in production engineering a product. A further advantage of the laser over electron beam is that the output from one laser may be fed to more than one welding station allowing one component to be welded while in the other station the next component is being loaded and prepared. With suitable optics and mirrors laser beams may be transferred over tens of metres (yards). Unlike the electron beam laser beams may be used with robots, the beams being passed down articulated arms to a welding head mounted on the robot.

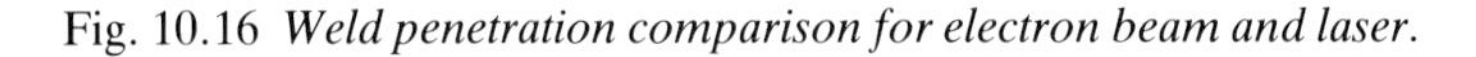

Fig. 10.16 *Weld penetration comparison for electron beam and laser.*

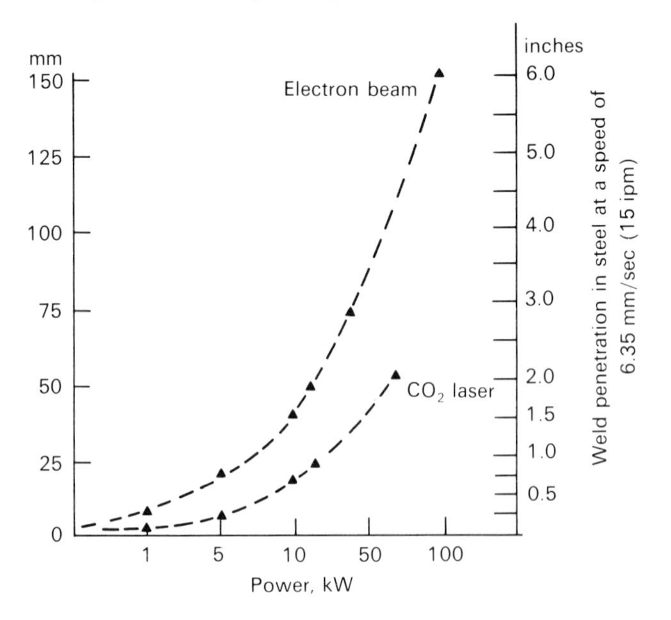

Process assessment

Power-beam welding and cutting with electron beam and laser, employ expensive and sophisticated equipment but the operating speed, quality and lack of distortion allow the processes to compete with arc welding in an increasing number of applications. Generally, there are no consumables such as filler wire and fluxes although there is a need to replace the gas in the laser on a continuous basis. Successful application does hinge, however, on the designer making maximum use of the advantages of the processes. The narrow deep welds with good contours, low overall heat input and distortion, make the processes precision assembly methods in a sense that other welding methods very rarely approach. Electron beam with its greater power capability is finding use in the welding of heavy material but the main areas of application at the present time are aircraft engine, aerospace, instrumentation and automotive industries.

Chapter 11

Robots for arc welding

In many of the earlier chapters reference was made to mechanised and automatic welding. By automatic welding was meant the use of a piece of equipment dedicated to a particular application which performed its task without continuous monitoring by the welder although the setting of welding conditions and the loading and unloading of parts would be performed manually. Mechanised welding is regarded increasingly as an intermediate technology which involves equipment that assists the welder but still requires that he sets and monitors the welding operation fairly continuously to maintain quality and productivity. The equipment for mechanised welding includes welding tractors, and various forms of manipulator, such as rotary fixtures. Automated operation is the highest level of working in which the complete operation including the mechanical handling is carried out without constant observation and adjustment of the welding parameters.

In its most advanced form automated welding makes use of devices for functions such as seam finding and seam tracking. These are necessary to take account of deviations from the anticipated path because of inaccuracies in the presented parts or the movement of the weld seam because of distortion introduced by the welding process itself. Seam finding and tracking systems require sensors which may be either contact types (such as probes, guide wheels or rolls) or non-contact types (such as electrical techniques based on the arc, electromagnetic or ultrasonic systems or vision systems employing structured light or laser beams and CCD, charge coupled diode, and video cameras).

Automation could not have developed without the microcomputer which has provided flexible control techniques and the ability to handle large amounts of information. Automation could not have been applied to arc welding without the development of electronic power sources which provide control of arcing conditions and allow for rapid changes in these conditions and also give stability against mains voltage fluctuations. The older automatic welding plants had electromechanical controls which once built were inflexible. Microcomputer control allows the welding plant to be programmed and reprogrammed at will to carry out a range of welding tasks which can also be linked in with other manufacturing activities.

The flexibility made possible by the microcomputer can only be deployed if there is also flexibility in the welding plant. This has been achieved in two ways, by (a) providing pivoting, sliding and jointed arms carrying the welding equipment, and (b) integrating the equipment for positioning the workpiece into the system to provide added degrees of freedom. This is the technology of robotics.

The basic unit

Hardware

The first robots were used for such applications as loading and unloading other machines and paint spraying, where accuracy and repeatability of movement was less demanding than for arc welding. Spot-welding was the first welding application for robots because it involved only a relatively simple operation of positioning. Arc welding requires that the robot can trace out a continuous path with an accuracy better than that of the diameter of the electrode wire used in welding. Equipment with this capability now exists and a repetition accuracy of better than ± 0.2 mm (0.008 in) is obtainable.

Most robots have a basic configuration belonging to one of the two types illustrated in Fig. 11.1. A popular form of arc-welding robot of type (a) which has the capability of reaching into difficult positions uses an articulated arm as shown in Fig. 11.2. It has motion on six axes and can be linked with a further six axes on manipulators or fixtures. A key feature of all arc welding robots is the ability to rotate the welding gun by 360° or more on one or more axes. This is called the wrist action and is illustrated in Fig. 11.3. Both hydraulic and electrical drives have been

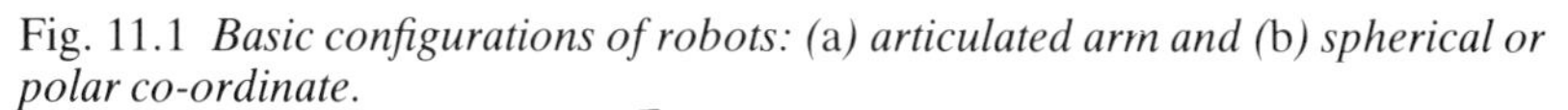

Fig. 11.1 *Basic configurations of robots:* (a) *articulated arm and* (b) *spherical or polar co-ordinate.*

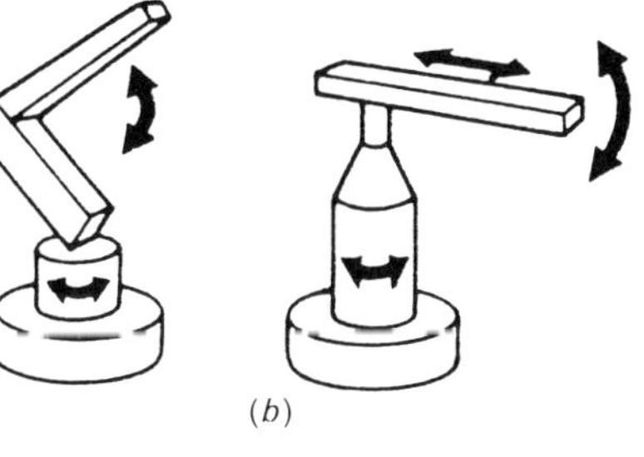

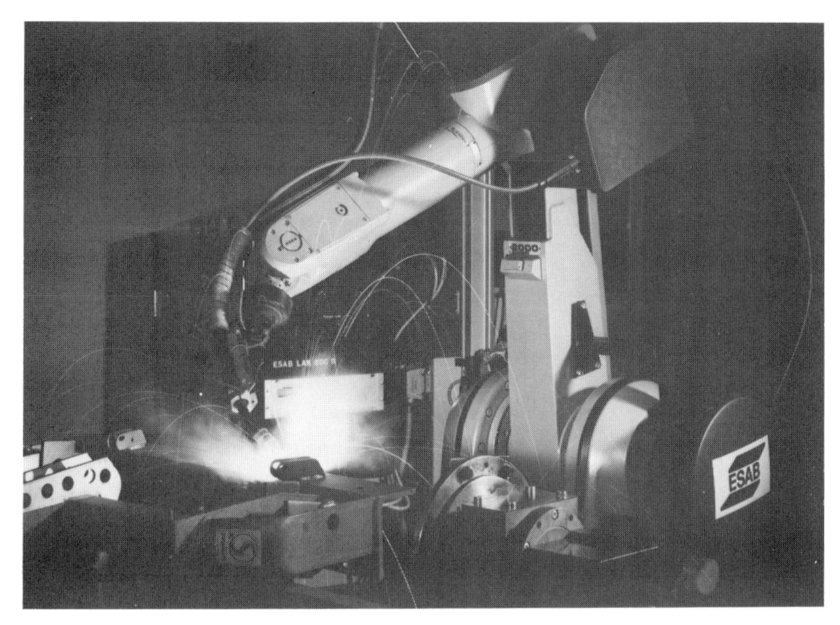

Fig. 11.2 *MIG welding robot of type* (a) *Fig. 11.1, welding small brackets on an assembly.*

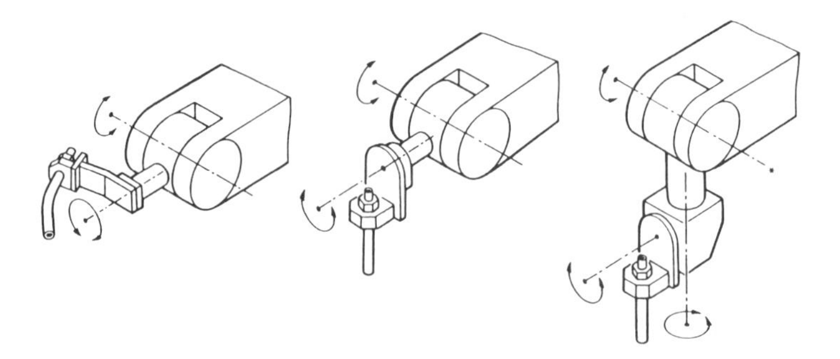

Fig. 11.3 *The final linkage in the robot arm to give 'wrist action'. The type shown at the extreme right gives an extra degree of freedom.*

used but electric drives are now the most common. They can be either DC or AC servo-controlled motors and must be able to drive the robot quickly and accurately and stop where needed without undesirable oscillation or back-lash. This usually means controlling speed in relation to the distance from the desired position which requires the use of tachometers to monitor velocity and encoders to monitor position. The control of acceleration and de-acceleration in this way allows faster positioning.

Programming

The computer control unit for the robot has a memory in which welding plant settings and instructions on sequencing and on the path it must follow are stored. A series of control buttons are arranged in a programming unit, sometimes called a teach pendant, which is connected by a cable to the computer control and may be held in the operator's hand. A system of 'teaching by showing' in which the torch, or a stylus replacing it, is moved through the desired trajectory can be employed on some robots. This trajectory will be stored in memory and reproduced on demand. More usually programming is carried out by driving the robot to a series of points along the desired path using push buttons on the teach pendant. These buttons have such instructions as: left, right, back, forward, up, down and various buttons to rotate the torch. At each point when the desired position has been obtained another button allows the instructions for that point to be recorded in the memory. The control software enables the robot to interpolate the trajectory between each recorded point giving a smooth path for the welding head to follow. There is a provision on some robots for a joystick control which speeds up the positioning of the robot on each point of the desired path. Once in memory the welding conditions, sequencing and weld path can be identified with the part being welded so that they may be recalled whenever the same part has to be welded. Figure 11.4 shows a teach pendant in use programming the part being

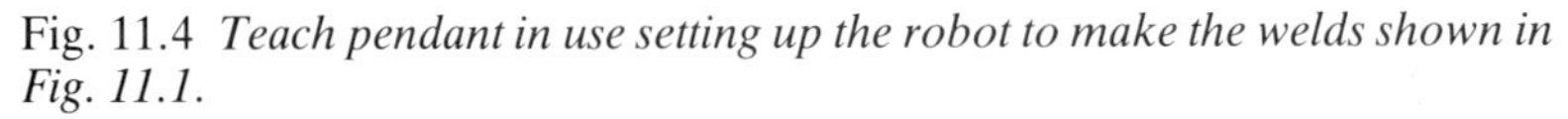

Fig. 11.4 *Teach pendant in use setting up the robot to make the welds shown in Fig. 11.1.*

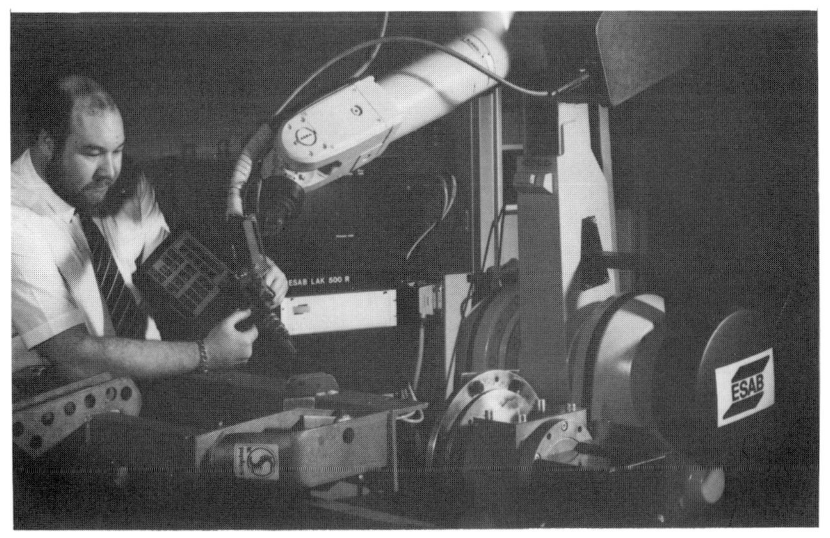

welded in Fig. 11.2. Although programming is a relatively slow and tedious exercise it has to be done only once for each workpiece. However, it does lock up the production equipment, and the development of off-line programming would be a considerable step forward. Techniques for off-line programming exist although at present they are rudimentary and must be developed further.

Fixtures

A robot can only achieve its full potential if it is teamed with a suitable manipulator. The manipulator moves the workpiece into the working area of the robot and will if necessary move the work continuously to place it always in the best position for the robot. Part of this equipment may be an indexing table which is necessary when a series of welds have to be made in different places on a workpiece. Standard welding positioners are generally unsuitable for use with robots, being slow and lacking the facility to be brought quickly to the desired position. A robot with a two-station indexing servo-controlled manipulator is shown in Fig. 11.5.

Working area

A robot as described above would typically cover a working area of about 1.5 m (60 in.) radius but this can be increased by mounting the

Fig. 11.5 *A two-station indexing servo-controlled manipulator teamed up with a robot.*

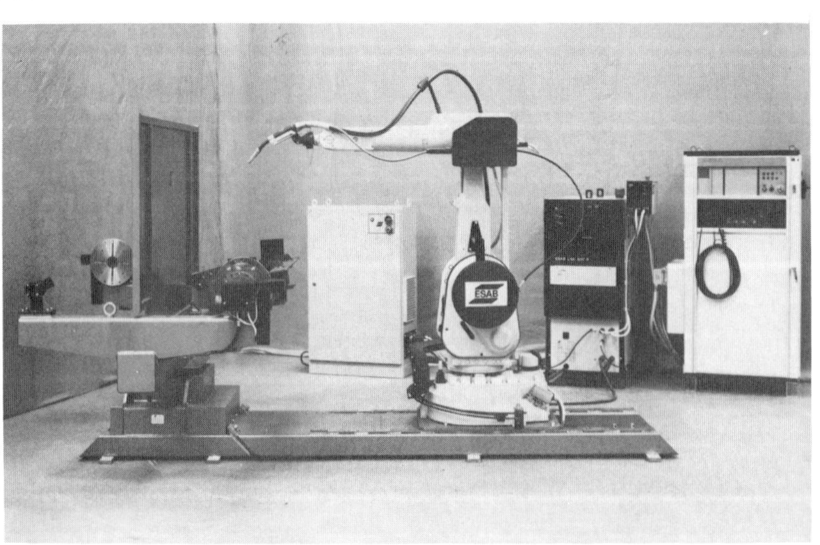

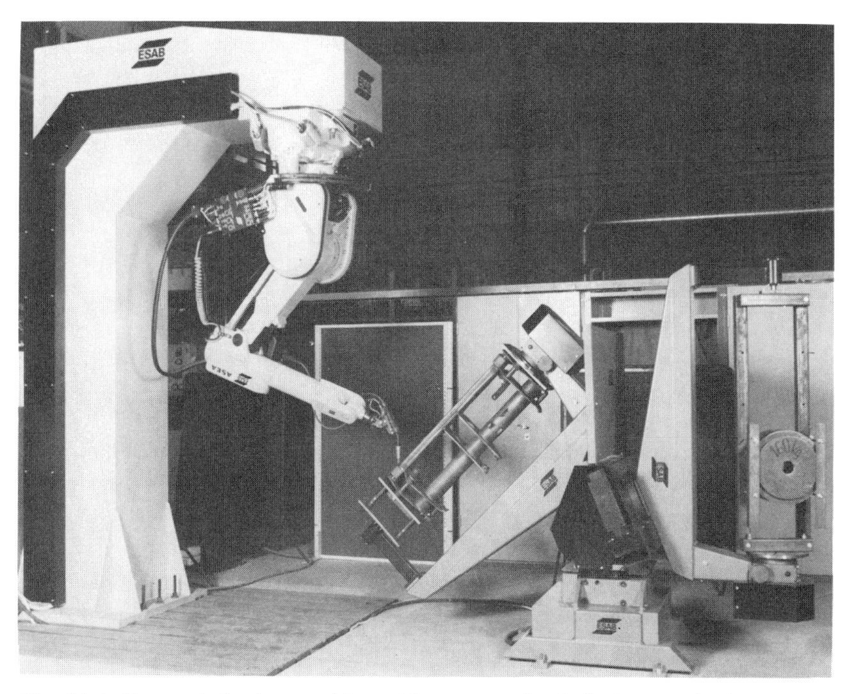

Fig. 11.6 *Suspended robot working with a two-station indexing manipulator.*

robot (or the manipulator) on a moving track. Where the robot must work at opposite ends of a long workpiece or move between two stations the robot may be mounted on an indexing track. To facilitate the movement of parts and keep the working area clear it is sometimes advantageous to hang the robot up-side-down from a column and boom or a gantry as shown in Fig. 11.6, and there is no reason why more than one robot should not be mounted on the same gantry.

Some companies offering robotic equipment have made a point of designing the various pieces of equipment mentioned in such a way that they are modular and can be assembled in various ways to obtain maximum flexibility. It is necessary, of course, that the computer controlling all the functions should have the required capacity and software.

The robot provides the opportunity to use arc welding in flexible manufacturing systems (FMS). In a typical example from the automotive industry one or more robots is suspended on a gantry above the working area, allowing the robots to work in a number of places. The work cell is served by an automatic unmanned truck which carries work to and from the various working points. Items to be welded are mounted on pallets and are placed in a buffer store. The computer controls all functions including queuing order and the calling up of the appropriate welding instructions when it has read the identity of the part on the pallets. As long as there is material in the queue for welding the cell can work without attention.

Joint finding and tracking

The application of robots has been greatly helped by the development of methods allowing the robot to find and follow the joint. These use a variety of principles and although some are well proved by practical use this is an area in which there is considerable experimentation and many systems, particularly those based on optical principles, require further development and practical trial. All methods appear to have limitations and must be applied intelligently.

Joint finding

Joint finding methods usually employ some part of the welding gun as a sensor, the gas nozzle, contact tube or the wire itself. The sensing part is energised with a voltage and the robot is put into a search routine. Every time contact is made with the work that point is registered and the robot continues to build up an image of the surface until if finds the shape corresponding to the start of the weld (see Fig. 11.7). The search routine and sensing voltage is then discontinued and welding is initiated. These systems can be a part of the through-arc joint tracking system to be discussed later. They are suitable for fillet, grooved and lapped joints.

Fig. 11.7 *Search routine for joint finding.*

Joint tracking

The earliest tracking systems were mechanical using wheels or rollers running in the joint itself. Electro-magnetic steering devices worked by a probe following the joint have been used successfully for long relatively straight seams. Robot welding, however, requires greater flexibility and the ability to track round tight corners. Probes making mechanical contact are also subject to a number of serious limitations, such as, a tendency to jump out of the joint and lose contact, they also limit the welding speed and are subject to wear. Contact tracking systems are not therefore suitable for robots.

Through-arc systems

Of the non-contact systems available the most widely applied at the present time is 'through-arc' sensing. It works by applying a weaving motion to the gun and monitoring the change in welding conditions as the electrode wire approaches the side walls of the fillet or groove (see Fig. 11.8). In this way the robot is able to find the centre of the joint. Vertical sensing or proximity control is carried out by monitoring the effect of electrode extension on welding current. Low electrical resistance electrode materials such as aluminium do not give satisfactory response in this respect. Through-arc tracking has the great merit of requiring no special equipment as it uses standard MIG guns. It is also a relatively low-cost system. The limitations are that the weld sidewalls must be well defined, the wire must feed straight from the contact tube without a cast or oscillation, the system is not fully effective on non-ferrous metals and the use of a weave is an essential requirement. Through-arc systems are marketed by several manufacturers, the main difference between them being in the microprocessors and software which manipulates the incoming signals.

Vision-systems

Numerous systems are under development employing straightforward or structured illumination, lens or fibre optics and ordinary video or CCD cameras. Probably the most important difference lies in whether the optics are external to the torch or contained within it, as if they are within the torch the system is truly 'real-time'. That is, the control signals are taken from the molten pool area itself and not from the joint ahead of the weld. However, sensing systems frequently employ laser illumination and CCD cameras and with the through-torch approach this equipment must be contained within the torch body. (The term torch is often used in the context of arc sensing although for the most part the systems are used on MIG guns.) There is possibly greater freedom in design when the sensing system is external to the welding

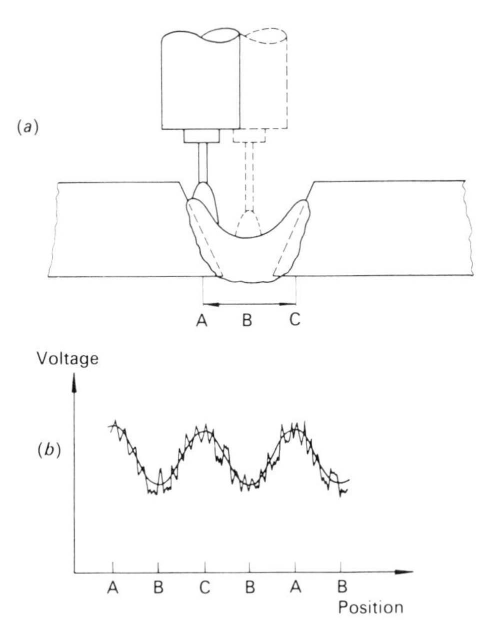

Fig. 11.8 *'Through-arc' sensing to provide seam tracking:* (a) *diagram of motion,* (b) *simplified voltage record indicating arc position.*

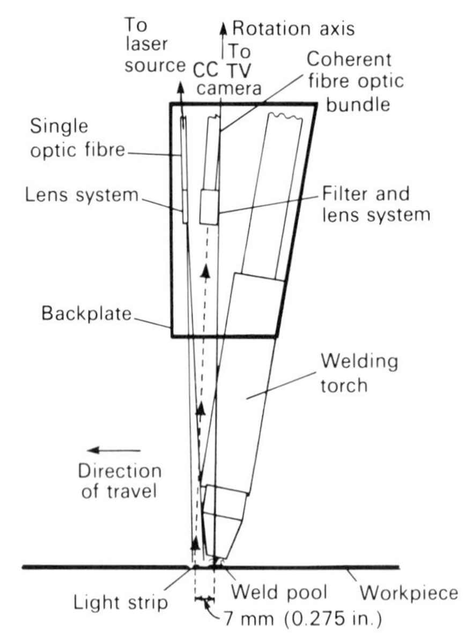

Fig. 11.9 *Diagram of a vision tracking system using optic fibres to connect with video camera and rotating facility to follow changes in direction (Liverpool University).*

gun. In this case, however, the sensing unit is taking information from the joint just ahead of the weld and when a change in direction is required the sensing unit must rotate around the welding gun. Figure 11.9 shows a TIG torch and sensing unit which works in this way, but the system is also applicable to the MIG process.

The illumination employed in joint sensing is frequently by Helium-Neon laser and is projected as what is often called structured light. That is, the laser beam is a narrow slice which forms a line when it falls across the workpiece. When viewed from an angle this line will reveal the contour of the surface and weld groove. The beam may be made into the slice form by scanning a spot to-and-fro across the joint. The function of the vision sensor is to provide three essential pieces of information; firstly, the proximity of the welding gun to the workpiece; secondly, the weld preparation contour to allow the weld centreline to be found and, thirdly, the shape of the joint path to help the robot track the weld seam.

Auxiliary equipment

Robot welding plants must work for long periods without attention and there are certain operations of a maintenance type which the robot may be designed to carry out itself. It may for example arrange for the nozzle of gas metal arc torches to be cleaned of spatter at pre-set intervals. The robot moves the welding gun to a position where a milling cutter and/or wire brush can be automatically inserted into the nozzle to remove spatter. It may also change a nozzle or even a complete torch. The latter may be necessary to allow a different size or composition of wire to be used in particular places in a complex assembly. Devices to find the start of a weld, or particular places in a workpiece, as well as seam tracking systems of varying degrees of sophistication are increasingly used as was mentioned earlier.

Process assessment

Although a robot cannot weld much faster than a human welder it does not tire or make mistakes so it continues to produce work of consistent quality. The main economic benefit of the robot is that it can weld continuously without meal and relaxation breaks and it is therefore much more productive overall. The flexibility of robot welding in a properly designed robot cell makes short production runs feasible. It is essential, however, to consider the whole production cycle if true economy is to be obtained. Work flow must be adapted to the robot system. An isolated robot will produce only modest advantages.

Although the usual arrangement is for the robot to carry the welding appliance it is possible to have the robot move the work under a stationary welding head.

Robots can work in conditions intolerable to humans. Not only are they insensitive to heat, noise, toxic gases, radiation and fume they can also work close together without problems from arc radiations. This gives scope in redesigning a production line. Most fusion welding robots use MIG or cored wire welding although plasma welding is also successful. TIG welding is more difficult because of the close tolerances which must be maintained but has been used with modern robots. Brazing by robot is also possible and presents no problems. Robots have also been used successfully in gas and plasma cutting.

The task of robot welding is easier with consistently accurate parts that are suitably jigged. Robots can, however, be made to use controlled weaving of varying degrees which increases the tolerance to inaccuracies in fitup.

Chapter 12

Welding productivity

Productivity may be measured in many ways. Where material control is the criteria it is often quoted as some ratio of output to input but to the fabricating industry it is more relevant to consider the time to produce a completed unit, or conversely the units produced in standard time. Important criteria are material flow, work station efficiency, fabrication (welding) speed and the level of rework. Which of these may become a major bottleneck or time-consumer depends on the particular company and items produced. In many fabrication shops, work stations and floor utilisation have developed over a number of years without regard to smooth material flow. Aspects such as siting of the material stock yard, the arrangement of stock material (e.g., is it as piles of steel plates or in 'toast racks'), crane capacity and heat treatment furnace size and position can totally dictate throughput time. These considerations are fundamental to a company wishing to remain competitive in an increasingly international and aggressive market place. Material flow is not the only factor, however, as even those fabricators with smooth lines and easy transportation may find that they fall behind because they are not using appropriate welding technology.

World best practice

Technological development now takes place internationally with increasing speed and companies are frequently compared on an international basis. Certain companies have become acknowledged as world leaders in their particular fields. They employ what has been called 'world best practice' in both management and engineering. In the more limited field of fabrication by welding it is noticeable that there is a considerable difference between fabricators in different countries particularly in their attitude to the use of robots and cored-wire welding. Certain leading international companies make extensive use of both and clearly do so because they are convinced that this gives them improved productivity. Other companies are still using the more traditional methods and it is necessary to ask why this should be. Perhaps the advantages are not always clearly seen and there are also a number of objections which may not have been examined thoroughly.

This chapter aims to demonstrate the increase in productivity and the effect on total performance which can be achieved by changing to reproducible automated welding with high-deposition-rate welding processes, and, as a particular example, the combination of robots and the cored-wire process.

High productivity creates opportunities

High productivity is not achieved simply by adopting advanced technology; that is only a tool which to be effective must be used properly. The primary requirements are a commitment to increased productivity taken at the highest level and a management programme to promote the policy through the entire organisation. There must also be a means for measuring and monitoring productivity improvements which must be seen to be of benefit to all levels of staff. Obviously it is preferable to aim for continued growth of output rather than workforce reduction and those companies investing most successfully in higher productivity welding solutions have used the productivity gains and the competitive edge this gives to expand their business.

A simplified example illustrates how this is achieved. Suppose a fabricating company has the capacity to manufacture 100 of a particular unit per year. The price charged for the goods must cover fixed costs (rent, rates, administration, etc.), variable costs (materials and labour), and profit. Assume that by adopting the best techniques this fabricator could double output to 200 units with the same labour force; this would give a lower labour cost per item and the ability to spread fixed costs over the additional 100 units as well. Although this gives more profit per item, it assumes that the additional units can be sold in the traditional way. The manufacturer could, however, be content with the performance in the established market place of the initial 100 units and continue to use them to cover the fixed costs. This would then allow the unit cost of the second hundred units to be reduced markedly and this benefit could be used to gain a share of an entirely new market. In this way the improved productivity can be used as a marketing tool, and the competitive edge can be used in a number of ways, e.g., to provide extra features for the same end price. Once the market is established the price can be raised to cover the cost of the extra features.

Objections to be overcome

A change of manufacturing process can bring problems to management and the whole workforce and to ensure a commitment to increasing productivity by investment it is necessary to meet the main objections to process change. These objections will vary according to the company and product but often stem from such factors as: a poor current trading position, bad experiences by other companies, concern over costs or the difficulty of making reliable cost predictions, fear of the imagined complexity of robots and a concern not to change too many things at the same time.

Depressed market position

A depressed market and poor order book can discourage investment in the latest technology. If the premises or their layout impede material flow, a large investment may be required to rectify this fundamental ill, indeed it may be too late to save such a site. But assuming that increased throughput could be accommodated, a reluctance to invest in welding improvements will not usually stand critical examination. Order books do not improve by themselves and with no investment to achieve improved productivity and lower unit costs the manufacturer who does not adopt the latest technology must inevitably fall further behind his competitors.

The poor experience of other companies

The unfortunate example of other companies who have failed in attempts to introduce a process change can discourage innovation. Such failures should be examined carefully, however, as they can often be attributed to a lack of appreciation of the critical factors necessary for success. Success depends to a large extent on the ready and enthusiastic acceptance of new processes by the workforce. The management programme must therefore cover the selection of the most suitable, not necessarily the cheapest, equipment after giving due consideration to performance, reliability, spares availability, service back-up and so on. An important technical decision with many processes is the choice of consumables. Other factors which have a direct bearing on the success of an installation include operator acceptance and training, and additional investment in ancillary equipment and services. It is pointless for example to install a robot which can fabricate components three times faster than a manual welder if material cannot be delivered to the work area any quicker.

Cost of consumables and plant

Cored wires are noted to be more costly weight-for-weight than conventional MMA electrodes or MIG wires and automation appears both costly and dauntingly complex to implement. Reassurance should be sought in the knowledge that since productivity is the ratio of value of goods produced over total cost of production it is the total cost of production which is important. The cost of consumables such as cored wires represents a small element, perhaps only 2 per cent of the total. What is costly is the capital plant, manpower and overheads needed for the production process. If the utilisation of these resources can be optimised the small additional cost of consumables should be recovered quickly. There are, however, frequently less obvious cost benefits to be gained as the example of automated cored-wire welding shows. The reproduceability of an automatic operation greatly reduces the risk of those defects commonly introduced by imperfect manual manipulation. The rectification of defects (when discovered) is often significant in terms of both the costs of rewelding and delays to completion. Undiscovered defects have other potentially more serious implications. Cored-wire welding gives consistent penetration and good fusion allowing with butt welds narrower edge preparations than can be used safely with MMA welding. Even ignoring the greater deposition rate of cored wires the reduced weld cross-section can improve productivity and reduce the usage of consumables.

A number of practical applications have been costed in detail and these have shown that the use of cored wire can substantially increase productivity and lead to better weld quality. Table 12.1 illustrates the effect of this change for 10 mm (0.39 in.) thickness vertical butt welds where it can be seen that the use of cored wire leads to a 110 per cent increase in output over MMA with a welding cost 43 per cent less.

Table 12.1 *Comparison of MMA and cored wire welding of a 10 mm (0.39 in.) thick vertical butt weld.*

		4 mm (5/32 in.) MMA	1.2 mm (3/64) mm FCAW
Electrode cost	£/hr ($/hr)	1.25 (2.2)	3.12 (5.46)
Deposition rate	kg/hr (lb/hr)	1.5 (3.3)	3.2 (7.05)
Duty cycle	%	20	20
Total welding cost	£/unit ($/unit)	32.20 (56.4)	18.53 (32.4)
Productivity	units/year	635	1355

NOTE: Labour £10/hr ($17.5/hr), i.e. direct cost only. Investment £1000 ($1750) for MMA and £3000 ($5250) for FCAW, 1800 working hours/year.

To put the case for cored wire in financial terms a balance sheet comparison has been made, using typical figures for small to medium companies operating in the UK, for a model company employing 25 workers. Welding costs, direct labour and consumables are taken as 10 per cent of total costs, fixed costs are set at 15 per cent and profit is taken as 10 per cent of sales value. The productivity figures from a computer-costing program for the single type of joint under consideration are then used to give the total cost, sales value, etc. The benefit from increased productivity is illustrated in Table 12.2 in two ways:

1. as a cost-cutting option where production is maintained and labour reduced, and
2. as a market share improvement option where labour is retained and output increased. The extra sales costs, e.g., increased discount, added features, are assumed to be 5 per cent of sales value.

While it can be seen from Table 12.2 that the cost-cutting approach, i.e., laying off excess welders, makes money, the profit is dramatically increased if advantage is taken of the increased output with a full labour force. Even allowing that sales costs may increase by 5 per cent of the selling price as a result of a policy either to increase market share in an

Table 12.2 *Comparison of FCAW and MMA for 10 mm (0.39 in.) thick vertical butt welding figures for costs in £K (£K, £1 = $1.75).*

	MMA	FCAW 'Cost cutting' option	FCAW 'Market share increase' option
Sales	5700 (9975)	5700 (9975)	12 162 (21 283)
Welding labour cost	476 (831)	223 (390)	476 (833)
Materials and consumables	37 (65)	75 (131)	160 (280)
Other variable costs	3762 (6584)	3762 (6584)	8026 (14 045)
Fixed costs	855 (1496)	855 (1496)	855 (1496)
Additional sales costs	—	—	646 (1130)
	5130 (8976)	4915 (8601)	10 163 (17 784)
Profit	570 (999)	785 (1374)	1999 (3499)
Improvement	—	215 (375)	1429 (2500)
Equipment cost		36 (63)	75 (131)
Redundancy		130 (228)	75 (131)
Total investment		166 (291)	75 (131)
Pay back time		9.2 months	2.8 weeks

existing market or to penetrate new geographical markets, the company makes £1.4m ($2.45m) extra compared with only £200 000 ($350,000) extra by taking a cost-cutting approach. Clearly selling the competitive edge can be used to great benefit and the pay-back for the investment to equip the 25 welders with cored-wire capability is actually less than three weeks.

Fear of robotics

Similar arguments can be made in favour of automation. The welding of 4 mm (0.16 in.) throat fillet welds is considered as an example. In the model company 25 manual welders are assumed to be using MIG welding with solid wire at 20 per cent duty cycle. Robots are easily capable of tripling this output by maintaining a 60 per cent arcing duty cycle (one operator per machine is assumed), so the productivity shown by the computer calculation in Table 12.3 is a simple threefold increase. If metal-cored wire is chosen, however, advantage may be taken of its ability to penetrate more deeply by using 3 mm (0.12 in.) instead of 4 mm (0.16 in.) throat fillets. The design strength is maintained as the total weld depth is still in excess of 4 mm (0.16 in.) The change in fillet size combined with the increased deposition rate results in over six times the productivity of the manual welder using solid wire.

It is impractical to make comparisons on the basis of one-for-one replacement of a manual welder by a robot plus operator. Flexible manufacturing centres with optimised material flow would make more sense than 25 individual robots. However, to compare solid and cored wire in a similar way to the previous example, where cost saving and market share growth approaches were considered, a combination solution is presented in Table 12.4. It is assumed that the 25-man

Table 12.3 *Comparison of manual MIG with automated solid wire and cored wire welding of 4 mm (0.16 in.) horizontal vertical fillets.*

	1.2 mm (3/64 in.) solid manual	1.2 mm (3/64 in.) solid automatic	1.2 mm (3/64 in.) metal cored automatic
Consumable cost £/kg ($/lb)	0.65 (0.51)	0.65 (0.51)	2.20 (1.75)
Deposition rate kg/hr (lb/hr)	6.2 (13.6)	6.2 (13.6)	8.3 (18.3)
Duty cycle, % total welding cost £/unit ($/unit)	1.43 (2.5)	1.00 (1.75)	0.62 (1.09)
Productivity units/yr	13 950	49 737	89 100

*£ = $1.75
NOTE: 3 mm (0.12 in) throat used with cored wire because greater penetration is possible.

Table 12.4 *Comparison of automatic and manual MIG for 4 mm (0.16 in.) throat fillet welds, costs in £K ($K, £1 = 1.75).*

	Manual	Automation + solid wire	Automation + cored wire
Sales	5700 (9975)	6709 (11 740)	13 676 (23 933)
Welding labour cost	425 (744)	170 (298)	170 (298)
Materials and	77 (135)	250 (438)	383 (670)
Other variable costs	3777 (6610)	4520 (7910)	9650 (16 888)
Fixed costs	855 (1496)	855 (1496)	855 (1496)
Total costs	5134 (8985)	5795 (10 142)	11 058 (19 352)
Profit	566 (990)	914 (1600)	2618 (4581)
Improvement		348 (610)	2052 (3591)
Equipment cost		800 (1400)	800 (1400)
Redundancy	—	150 (263)	150 (263)
Total investment		950 (1663)	950 (1663)
Pay back time		32.4 months	6 months

NOTE: 3 mm (1.2 in.) throat used with cored wire because greater penetration is possible.

workforce is reduced to 10 and that each is equipped with a robot. Table 12.4 shows that the increased output gives a payback when solid wire is used which is close to three years. If the cored-wire solution is adopted the pay-back period is six months.

When embarking on a programme of change to robot production it is even more important that the management action plan considers the overall package of equipment, data, job costing and service available from the supplier. Some fabricators have been led to believe that a low-cost robot with no back-up is the solution. Months of inactivity will usually demonstrate that this is a poor choice indeed.

Too much change

Converting to robot or mechanised welding techniques at the same time as making a change to cored-wire welding may be thought to be introducing too much new technology at the same time. Table 12.4, however, has already shown the powerful economic argument in favour of combining automation and cored-wire welding. Any change, however small, can be critical if handled badly but a company operating a committed management action plan can cope with any level of fabrication process change by working in co-operation with its suppliers and seeking help where needed. There are many examples of companies which have successfully made such changes to put themselves ahead of their competitors.

Appendix 1

Health and safety

Any industrial operation in a workshop or on site can carry an element of risk and safe working requires an assessment of risk and vigilance on the part of both employer and employee. Welding is no better or worse than many other industrial occupations but like all processes has a number of specific risks which must be appreciated. The responsibility for the provision of a safe and healthy working environment rests with the management and the workforce has a duty to co-operate with the management to achieve this end. The notes which follow are an introductory guide and users of welding are advised to obtain information from both the manufacturer or supplier and the sources listed at the end of this appendix.

It is emphasised that the starting point for a safe workplace is that it should be clean and orderly. The four main areas which have to be considered when welding are:

1. Electrical safety.
2. Heat, fire and explosion.
3. Light.
4. Fumes and gases.

Electrical safety

Having checked that the arc welding equipment has an appropriate rating, particularly duty cycle for the intended use, it should be installed in accordance with the manufacturer's instructions and any relevant British Standards. Terminals and live components should be properly protected and earthing circuits should be of adequate capacity. Note that there is a separate earthing conductor in addition to the welding current return cable. All cables and connectors should be of a size and construction suitable for the maximum current they are likely to carry.

Where possible arc welding equipment should be of a type allowing open-circuit voltage reduction. Isolation switches should be readily accessible and operators should know where they are.

All equipment, especially cables, electrode holders and torches should be regularly inspected for damage or wear which should be repaired immediately or the item replaced. Damage to cables can be minimised by avoiding unnecessarily long primary cables or welding cables trailing across workshop thoroughfares.

The danger from electric shock is increased in hot, wet or damp conditions. Skin contact with a welding electrode should be avoided by the use of dry welding gloves in good condition. The consequences of an electric shock can be more serious when working in confined spaces or at heights when there is a risk of falling.

Electron-beam and laser welding equipment use voltages of 20,000 and upward. The high voltage parts are well protected by the manufacturers and access panels should always be securely in place before the equipment is used. As with all electrical equipment, if accidentally subjected to damp or wet conditions the power should be immediately switched off and a competent person consulted on remedial measures.

Heat, fire and explosion

Appropriate protective clothing should be worn to guard against, heat and hot particles when welding or cutting. Particularly when welding or cutting above bench height there should be no recesses in clothing or footwear in which hot particles may become lodged. Clothing should be free from grease and preferably made of wool which is not so readily flammable as synthetics. Metal which is hot should be guarded and clearly marked.

No welding should be carried out in the vicinity of flammable vapours or combustible materials. When welding or cutting above ground it should be remembered that sparks and hot metal can travel for a considerable distance. Sparks from cutting can travel up to 9 m along a floor. Fire-fighting equipment should be readily available and its position known to the operators.

Cylinders for gas welding and cutting should never be used when they are lying in the horizontal position. When in use they should be fastened securely in the vertical position to a wall or stanchion or to an approved trolley. Regulators should only be used for the gases for which they were designed and they should be treated with care and not for example exposed to rapid pressure surges by the sudden opening of cylinder valves. The pressure on the control screw should be relaxed after the cylinders have been closed and pressure in the hoses has been released when shutting down after welding.

Hoses should be checked regularly and joints made with approved connectors, never copper couplings with acetylene as this could cause the formation of the potentially explosive copper acetylide. Fitted hoses are best, especially those having safety check valves to prevent back-feeding. Flash-back arresters should be used. A leak of gas may often be detected by a hissing sound or, with fuel gases, by smell. Leaks should be located with a solution of Teepol (detergent) in water and then investigated and rectified.

The welding or cutting of drums, containers or tanks which have held flammable liquids or gases can be hazardous even when supposedly clean. The welding equipment manufacturers or a recognised authority should be consulted for precautions which must be taken. When vessels or hollow parts are to be welded it is necessary to ensure that these are vented to prevent build-up of pressure. Oxygen must not be used for blowing out or cleaning pipes or other equipment.

Light

Apart from visible light welding operations often emit both ultraviolet and infrared radiation against which the operator and those working nearby must be protected. In gas welding and cutting operations little ultraviolet is emitted and protection is required mainly from the extreme brightness of the pool and infrared which can damage the cornea of the eye. For this purpose goggles are effective. When arc welding there is an additional requirement for protection from ultra-violet emissions. Goggles are quite unsuitable for use with arc welding because they offer no protection to exposed skin and the filters are not sufficiently dark. The amount of UV emitted from an arc depends on the welding current and as this is raised the filters through which the arc are observed must become darker. British Standard 679, 'Filters for use during welding and similar industrial operations', gives information on the density required for various operations and ISO standards 4849–4852 cover the corresponding area. In the USA, the relevant specifi-cations, are ANSI/AWS Z49.1, 1983, 'Safety in welding and cutting', Z87.1 'Practice for occupational and educational eye and face protec-tion' and AWS F2.2, 1984, 'Lens shade selector'. Filters for arc welding are mounted in hand shields or better in helmets and are protected on the outside from spatter by a separate plain glass. Helmets and hand shields should be in good condition and the grade of filter should be checked to ensure its suitability before use.

Accidental exposure to the light from an arc can cause a condition known as 'arc-eye'. This painful condition does not develop immediately but can persist for a day or two before disappearing. Persistent or prolonged exposure without adequate protection can, however, result in more serious damage. Personnel working near a welder are also at risk even when not facing directly to the arc. Suitable safety spectacles should be worn by all personnel entering a welding area. Notices warning of arc flashes should be displayed at the entrance to the screened-off welding areas. Welders should always warn others nearby before striking an arc.

Glare can be reduced by using non-reflecting surfaces to walls and screens and by avoiding white clothing. Gas-shielded processes have to be treated with more care than manual metal arc used at the same welding current where the arc is slightly screened by flux. Polished sheet workpieces, e.g., aluminium, are also good reflectors. Ultraviolet light can cause skin inflammation similar to sunburn and appropriate dark coloured protective clothing should be worn.

Fumes and gases

Many welding processes give rise to fumes. The scope for reducing or changing the composition of fumes by changing the composition of electrodes is extremely limited. Management must ensure that problems are identified through proper monitoring and analysis and, where appropriate, that adequate equipment is provided to capture and treat fume. The work force must be trained in the proper use of this equipment and must ensure that it is used when necessary.

Welding fumes consist of various airborne substances in the form of gases or fine particles which may create hazards to health when inhaled or swallowed. The degree of hazard to the welder depends on the composition of the fumes, their concentration in the air breathed and the time of exposure. The effects of excessive exposure to fume arising from inadequate ventilation may become apparent at the time of welding or at some later date.

Welding fume characteristics

Particulate welding fume is generated at the welding point and is seen as a rising column of finely divided particles above the welding arc. The particles are mostly oxides and silicates of metals which have volatilised during transfer across the welding arc. The gases nitric oxide, nitrogen dioxide and ozone may sometimes be produced by the action of the electric arc or the radiation from it on the surrounding air. These gases do not arise from the welding consumables and are not usually a problem in MMA welding under conditions of normal ventilation. MIG welding is more likely to give rise to these gases particularly at high current levels and ozone generation may be increased by the presence of argon in the atmosphere around the arc. The light from the welding arc can cause the vapour from degreasing operations to break down into noxious gases even at some distance from the arc. Carbon monoxide may be produced by decomposition of carbon dioxide in the shielding gas or of carbonates in flux-cored wires.

Factors affecting exposure

The composition and concentration of fumes depend on:

1. The process, the type and size of electrode, the duty cycle and the current and voltage used.
2. The dispersal of the fumes – by natural air movement or mechanical ventilation and the effectiveness of local extraction employed.
3. How close the welder is to the arc, and the position of his head in relation to the rising column of fume.
4. The arrangement of the workpiece and degree of confinement of the working space.

Hazards of excessive exposure to fume

Some of the effects of fume are summarised below:

Irritation of the respiratory tract This is the effect of dust or fume on the lining of the respiratory tract and can cause dryness of the throat, tickling, coughing, chest tightness, wheezing and difficulty in breathing. In its most acute form, as with cadmium fume, it can cause the lung to become full of fluid. The effects will vary with exposure, concentration and type of irritant.

Metal fume fever The inhalation of many freshly formed metallic oxides such as those of zinc, cadmium, chromium, nickel, copper and manganese may lead to an acute influenza-like illness termed 'metal fume fever'.

Systemic poisoning This can result from the inhalation or swallowing of substances such as fluorides, hexavalent chromium, cadmium, lead and barium.

Long-term effects It is possible that certain constituents of welding fume such as hexavalent chromium and nickel may be carcinogenic and until there is definite information about this it is wise to treat them as such.

Fibrosis This is the formation of fibrous or scar tissue in the lung and it is the result of a reaction between dust or fume with the lung tissue. There are various types depending on the nature of the substance involved and duration of exposure.

In all cases of doubt concerning physiological response to welding pollutants, medical advice should be sought promptly.

Occupational exposure limits

The recommended limit on the concentration of welding fume (or any other atmospheric contaminant) in the air breathed is defined by the Health and Safety Executive in a list of Occupational Exposure Limits (OEL), Guidance Note EH40. This guidance note is revised annually and reference should always be made to the most recent edition. A long-term exposure limit (8-hour TWA value) of 5 mg/m^3 for particulate welding fume is included in the current OEL Recommended Limits list.

It is the responsibility of the user/employer under the Health and Safety at Work Act that OELs are not exceeded. The fume analysis cannot be used to assess the concentration of total welding fume to which a welder is exposed. Assessment of the possible exposure of the welder must be carried out by a competent person and may involve air concentrate measurements in the workplace. The analysis of fume from electrodes and wires for welding mild and some low alloy steels and aluminium alloys indicates that at a total particulate fume concentration of 5 mg/m^3 no individual constituent of the fume will exceed its own recommended limit. However, there are consumables, e.g., those for stainless steels, hardfacing or copper welding, which give fume containing elements such as chromium, nickel, manganese and copper in sufficient quantity that even at 5 mg/m^3 their own limits would be exceeded. A greater degree of fume control or protection is therefore required to ensure that welders and other are not exposed to excessive amounts of these elements. Special care may also be required when welding coated, plated, painted or primed work.

Control of exposure

The manufacturer's recommended welding conditions should be used and it should be noted that excessive current generally produces more fume without improving metal deposition. The welder's head should be kept out of the fume column. Adequate ventilation based on the welding duty cycle used in making the original hazard assessment should be provided to ensure that the level of welding fume in the operator's breathing zone is less than 5 mg/m^3. The level of ventilation provided should also ensure that any relevant individual occupational exposure limits are not exceeded.

Appropriate precautions may range from good general ventilation to a high standard of local fume extraction applied as close as practicable to the source of the fume which is invariably the most effective form of fume extraction. When working in confined spaces the provision of airline breathing apparatus should be considered. Portable extraction fans or self-contained extractor/filters can be used when a fixed installation cannot be employed.

If in doubt as to the precautions to take reference should be made to the equipment or consumable supplier, The Welding Manufacturers' Association, 8 Leicester Street, London WC2H 7EN or The Welding Institute, Abington, Cambridge CB1 6AL. Information from The Health and Safety Executive may be obtained through HMSO, London.

Appendix 2

Glossary of terms

The following list is a selection from the terms commonly used in the context of welding and cutting. The definitions given do not always follow exactly those in BS 499: Part 1: 1983, containing over 2000 definitions but the compilers wish to acknowledge the excellence of this document as a basic source. A number of the terms given below do not appear in BS 499 but the definitions follow accepted usage.

There are frequently differences between the UK standard, BS 499: Part 1: 'Welding Terms and Symbols', and the USA 'Standard Welding Terms and Definitions' drafted by the American Welding Society (AWS). These differences are often important and were highlighted in Section 8 of Fabrication 86 a publication of The Welding Institute, Abington, Cambridge, UK, with whose permission this glossary is reproduced.

Acid Flux An acid covering for a metal arc electrode is one containing a large proportion of iron oxide. Such electrodes are now little used. Acid fluxes for submerged arc welding have a high proportion of silica.

Agglomerated flux A flux for submerged arc welding made by mixing the finely ground constituents with an aqueous solution of a binder such as sodium silicate followed by drying and baking. Submerged arc flux is often called 'the melt'.

Air plasma cutting Plasma cutting in which it is possible to use air instead of an inert gas because the electrode is not tungsten but a material resistant to oxidation.

Arc-air cutting Thermal cutting in which an arc, usually from a carbon electrode, melts the metal which is removed by a blast of compressed air.

Arc blow Deflection of the arc by magnetic fields in the surrounding material.

Arc-spot welding The welding of overlapping workpieces by a stationary arc which forms a molten pool penetrating the upper part and fusing into the lower part.

Backing technique A technique to control a weld bead which penetrates through to the underside of a joint. The backing may be permanent, as when an additional piece of metal is fused into the joint, or temporary as when a strip of metal or other material is positioned to support the penetrating bead until it solidifies.

Basic electrode An electrode having in its covering a high proportion of calcium carbonate and fluoride.

Basic flux A flux for submerged arc welding having as major constituents calcium fluoride and oxides of calcium, magnesium and manganese. Can be applied to the covering of a manual metal arc electrode.

Blowpipe Usually a hand-held burner which mixes the gases and produces a flame suitable for welding, brazing and cutting.

Bore welding Tube/tube plate welding or tube butt welding using an automatic tungsten arc welding device which operates from inside the tube.

Braze welding Joining using a technique like fusion welding but with a filler metal having a lower melting-point than the parent metal which is not melted intentionally.

Brazing A joining process in which under the action of heat a filler metal with a lower melting-point than the parent metal melts to fill by capillary attraction the space between the parts to be joined. When the filler metal has a melting-point below 450C (840F) the technique is known as soldering. The filler metal may be pre-placed and the heat may be supplied by flame (flame brazing) by immersion in a bath of molten brazing alloy (dip brazing) by immersion in a molten salt bath (salt bath or flux-dip brazing) by heat from high-frequency induction (induction brazing) or a furnace (furnace brazing).

Bulk welding A proprietary form of submerged arc welding or surfacing in which granulated metal is fed into the joint ahead of the flux to become incorporated into the weld metal.

Burnback The fusing of the electrode wire to the contact tube in gas metal arc or submerged arc welding as a result of a sudden accidental increase in arc length.

Burnoff rate The rate at which a consumable electrode is consumed in an arc. May also be applied in friction and flash welding to the rate at which the workpieces are shortened.

Butt weld A complicated definition is necessary for both a butt weld and a butt joint and both terms are used indiscriminately. It is best to consult BS 499, or AWS standard; however, there are important differences between these standards.

Buttering The deposition on one or both fusion faces of a joint of a layer of weld metal which may have a different composition from the parent metal, to provide an intermediate or 'buffer layer'.

Capping pass The final pass of a multipass weld especially applied in pipe welding.

Cellulosic electrode An electrode with a covering having a high proportion of cellulose.

CO_2 welding Gas metal arc welding using carbon dioxide as the shielding gas.

Conduction limited weld A weld made by electron beam, laser or plasma welding at a power intensity insufficient to cause keyholing, the heat being transmitted through the thickness of the material by conduction to form a weld of bowl-shaped cross section.

Constant current power source A power source the current output of which is little affected by changes in arc voltage (see drooping characteristic).

Constant potential power source A power source the current output of which shows marked changes for small changes in arc voltage (see flat characteristic).

Consumables All supplies consumed in the welding operation but particularly electrodes, wires, fluxes and gases.

Consumable electrode An electrode that provides filler metal, but generally a continuous wire electrode either solid or cored.

Contact tube A copper tube through which a continuous wire electrode passes to pick up welding current and be guided.

Core wire The metal wire covered by flux which forms a manual metal-arc welding electrode.

Cored wire A continuous electrode usually made from strip formed round a core of flux. Hence the use of the term flux-cored wire but metal-cored wires (flux and metal powder core) also exist.

Crater The depression formed on the last metal to solidify at the end of a weld.

Deposited metal The metal deposited from an electrode or filler wire, including any iron powder in the flux fused into the weld metal.

Deposition efficiency An AWS definition, also known as metal recovery or electrode efficiency (BSI definition). The mass of metal deposited from a metal arc electrode expressed as a percentage of the mass of the core wire, excluding the stub end. With iron-powder electrodes metal from the covering results in percentages exceeding 100.

Deposition rate The mass of metal deposited in unit time, often expressed as kg/hr.

Diffusion welding Sometimes called diffusion bonding, a joining process in which under the action of heat and pressure union across the faying surfaces is achieved by atomic diffusion without melting or serious plastic deformation.

Dilution The change in composition of the metal deposited from a filler wire or electrode as a result of mixing with the melted parent metal. Expressed as a percentage of the melted parent metal in the weld metal.

Dip brazing, dip soldering See under Brazing.

Dip transfer The form of metal transfer from a consumable electrode in which metal is detached as a result of a rapid succession of short circuits between the electrode and the molten pool. Also known as short-arc, particularly in the United States of America.

Drooping characteristics The volt-ampere characteristic of a constant-current power source.

EB insert A backing technique for pipe welding originated by the Electric-Boat Company, USA, in which a permanent backing ring is fused into the weld.

Edge joint A joint in which the edges of the parts to be joined meet at an acute angle. See BS 499 for closer definition.

Electrode In welding, a solid conductor through which an electric current enters or leaves the welding arc or the medium in which heat is generated. The metal rod with a flux covering which is fused in the manual metal arc process sometimes called a stick electrode; or the bare wire fused by an arc under powdered flux (submerged arc) or a gas shield (MIG or MAG); or a non-consumable tungsten rod and a gas shield (TIG); or the copper alloy rods clamping the workpiece in resistance spot or seam welding to convey the current for resistance heating.

Electrode force The force transmitted by resistance welding electrodes to the workpiece.

Electrode holder The device which holds a metal arc electrode and conveys current to it. In resistance welding the device which holds an electrode is called a bolster.

Electrode pick-up The contamination on the surface of a resistance welding electrode from the metal being welded or scale or coatings on the metal.

Electrogas welding Arc welding using a gas-shielded consumable electrode which is carried out vertically using cooled shoes to retain the molten pool in position.

Electron beam Welding with a focused beam of electrons. Normally carried out at low pressure in a chamber but can be done in open air when it is called non-vacuum electron beam welding.

Electro-slag welding Vertical fusion welding in which one or more electrodes is fused in a bath of molten electrically-conducting slag held in position by cooled shoes. The shoes may be advanced to keep pace with the weld if it is too long for a single pair of shoes to be used.

Excess weld metal The term preferred in the United Kingdom to reinforcement for weld metal lying outside the plane joining the weld toes. In the United States of America, reinforcement is the standard term.

Fade-out region In a circumferential electron beam, laser or arc weld the region past the overlap in which welding power is progressively reduced to zero prior to discontinuing welding.

Faying surface The surface of one component that is intended to be in contact with a surface of another component to form a joint, e.g., the overlap in a lap joint.

Filler metal Metal added during welding, brazing or surfacing which may be in the form of wire, rod or strip. When added prior to the welding or brazing operation it is 'preplaced'. In the United Kingdom, filler wire means filler metal in wire form but in the United States of America, welding wire is the normal term.

Fillet weld A fusion weld other than a butt, edge or spot weld which is roughly triangular in transverse cross section and joins two members at an angle to each other. The size of a fillet weld is defined in part by the term leg length. In the United States of America, a full 'fillet' is one having a size equal to the thickness of the thinner member being joined.

Filling pass The weld passes in a multi-run weld intermediate between root and capping passes.

Flash Previously the term used to describe the fin of metal extruded from the joint as a result of forging. BS 499 prefers 'upset metal' and applies flash to the spatter expelled during the flash welding process.

Flash welding A resistance butt welding process in which the two parts of the workpiece are advanced together in such a manner that melting and local arcing can occur to generate heat at the faying surfaces prior to their being forged together.

Flat characteristic The volt-ampere characteristic of a constant-voltage power source.

Flat position See Fig. 4.5(*a*) and (*b*) on page 62.

Flux A substance placed on the surface of the workpiece, electrode or filler wire which melts during welding or brazing to clean the surfaces and protect hot metal from atmospheric contamination. It may also be used to make additions of deoxidants and alloying elements to the molten pool.

Flux-cored arc welding (FCAW) See Cored wire.

Friction welding Welding in which heat generated by friction from relative motion of the faying surfaces is used to make them plastic so that they may be united by the application of pressure during or after the arrest of motion. The most common form of friction welding is for welding bar or tubular parts one of which is stationary, the other being rotated in contact with it under pressure. Two mechanical arrangements are known, 'continuous-drive', which uses power drive through the welding cycle, and 'inertia' in which energy is supplied from a flywheel.

Friction welding – radial Friction welding using frictional heat to render a revolving ring of metal plastic which is then collapsed between stationary members, e.g. pipe ends, or on to a cylindrical surface to make a flange.

Fused flux A flux for submerged-arc welding made by melting the constituents together and crushing the product. See also agglomerated flux and sintered flux.

Fusion face The portion of a surface, or of an edge, that is to be fused in making a fusion weld.

Fusion welding Welding in which parts are unified by metal which has been melted, e.g., by arc, flame or power beam.

Fusion zone the part of the parent metal which is melted into the weld metal.

Gas metal arc (MIG) welding Gas-shielded arc welding with a continuous consumable electrode shielded by inert or other gases.

Gas-shielded arc welding Arc welding in which protection from atmospheric contamination is achieved by a flowing gaseous shield, usually argon, helium or carbon dioxide or mixtures of these.

Gas tungsten arc (TIG) welding Gas-shielded arc welding in which the electrode is non-consumable tungsten and the shielding gas argon or helium.

Gas welding Fusion welding using the heat from a flame, usually acetylene burning in oxygen. Oxy-fuel gas welding in the United States of America.

Globular transfer The form of transfer from a consumable electrode which comprises globules with a diameter greater than that of the electrode and which are transferred largely by gravity.

Gravity position Flat position.

Gravity welding A simple form of mechanised welding employing a metal arc electrode of the 'touch' type in which the electrode holder is incorporated in a slide or hinge mechanism allowing it to be both held in contact by gravity with the workpiece and moved along the joint. In its basic form a tripod with one leg the electrode.

Groove The term used in the United States of America for a prepared joint. Used in defining welding position.

Guide tube A rigid tube that guides but does not convey current to a filler wire or electrode.

Heat affected zone (HAZ) That part of the parent metal which is affected by the heat of welding or cutting but not melted.

Heat control The smooth control of current output from a power source by electronic means, usually phase shift control of thyristors.

Horizontal-vertical welding The BS term for making a weld horizontally in plates one or both of which is substantially vertical. See below for American terminology. See also Fig. 4.5 (*a*) and (*b*) on page 62.

Horizontal welding The AWS term for the welding position in which the weld is horizontal. Equivalent in BS 499 is horizontal-vertical. Compare with flat position.

Hot-wire TIG Tungsten arc welding in which a filler wire fed into the molten pool carries current from a supplementary power source thereby increasing the melting rate of the wire and deposition rate of the system. See also I^2R heating.

Induction brazing Brazing in which the heat is supplied by inducing a high-frequency current in the workpiece.

Induction welding A welding process often applied to seam welding tubes or other continuous processes in which an induced high-frequency current fuses the surfaces of the joint prior to their being forged together.

Inertia welding A friction welding process in which the mechanical energy is obtained from a flywheel (US term).

I^2R heating Heat generated by electric resistance but a particular meaning is the heating of a continuous wire electrode in gas metal-arc or submerged-arc welding by the arcing current. The stickout may be deliberately increased to enhance the effect and raise the burnoff rate and hence deposition rate.

Intermittent welding A series of welds of the same type and dimensions at intervals along a joint.

Inverter power source The conversion of low-frequency AC to high-frequency AC by rectifying the low-frequency current and chopping the DC so produced. In arc welding power sources the conversion of mains voltage AC to a high frequency allows advantage to be taken of the fact that high-frequency transformers are substantially smaller than their low-frequency equivalents. After the voltage has been reduced to a value suitable for welding the high-frequency current is rectified to provide a DC output for welding from an extremely compact portable power source.

Iron powder electrode A metal arc electrode having in its covering an appreciable amount of iron powder, generally sufficient to give a deposition efficiency of over 130 per cent.

Keyholing The effect which makes possible the deep, narrow welds using high-intensity heat sources such as electron beam, laser and plasma. A hole is produced through the workpiece by melting and evaporation, this being carried forward to make the weld.

Lack-of-fusion Failure to achieve fusion between weld runs (lack of inter-run fusion) or between weld and parent metal (lack of side-fusion) or between parent metal and parent metal (incomplete penetration).

Lamellar tearing Separation through planes of slag inclusions in heavy steel plate adjacent to welds and associated with cracking.

Lap joint A joint between overlapping parts which are substantially in the same plane. Completed by fillet, spot or plug welds.

Leg length A measurement of fillet weld size being the distance from the actual or projected intersection of the fusion faces to the toe of the weld.

Low hydrogen electrode An electrode normally of the basic type which is capable of producing steel weld metal with a diffusable hydrogen content of less than 15 ml/100 gm (15 cc/100 gm).

Magnetic amplifier control Control of welding current by a choke (reactor) having an additional coil carrying DC the current in which can be varied. Sometimes called a saturable reactor control.

Mag Welding Gas metal-arc welding with the active gas carbon dioxide.

Manipulator A mechanical device manually or power operated to bring or continuously rotate work to the desired position for welding. The word, positioner, now deprecated was previously used as a synonym but would be best used to indicate a device used to bring work into the desired position for welding. A manipulator would then be a device to continuously move the work to keep it in the desired position.

Manual metal arc (MMA) Arc welding with a straight covered electrode of suitable length manipulated entirely by the operator. Known as shielded metal arc welding in the United States of America.

Melt run Parent metal melted by the passage of a heat source in simulation of a close square butt joint. Compare with bead-on-plate for an arc weld with a consumable electrode.

Metal-arc welding Arc welding with a consumable electrode.

MIAB welding Magnetically impelled arc butt welding, in which for the purpose of joining thin walled tubes or hollow sections, the faying surfaces of the workpieces are melted by an arc impelled around the weld line by magnetic fields, the joint being consolidated by a forge.

MIG welding Gas metal-arc welding with a bare wire and a gaseous shield.

Motor generator A combination of an electric motor and generator to supply DC for welding.

Multiple electrode welding Arc welding with more than one electrode usually used in the context of submerged arc welding.

One-sided welding Full penetration welding completed wholly from one side of a workpiece often using a backing technique.

Orbital welding Mechanised welding of pipes or tubes in which the welding head moves round the outside of the stationary workpiece.

Penetration The depth to which in fusion welding that the parent metal or previous weld runs has been fused. In spot welding the depth to which the nugget penetrates each sheet.

Phase shift control See Heat control.

Plasma cutting Thermal cutting in which a constricted tungsten arc with extra concentric gas shielding melts a narrow region of the workpiece, the melted metal being ejected by the arc and gas force. See also air-plasma cutting.

Plasma welding Low-current plasma, less than about 150 A, and micro-plasma, less than 10 A, uses a constricted tungsten arc for making normal conduction limited welds. High-current plasma up to 400 A is used for welding metal with the keyhole technique.

Plug weld A weld made by filling a hole in the upper member of a lap joint so as to fuse into the surface of the underlying member.

Polarity Because of a transatlantic confusion in the terms straight and reverse polarity these terms are to be avoided. Polarity is now indicated by reference to the electrode, i.e. either DCEP or DCEN.

Positional welding A term with some transatlantic confusion. Best to refer to welding position directly, i.e, flat, horizontal-vertical, vertical or overhead. See BS 499 or AWS 'Standard Welding Terms' for exact limits, also Fig. 4.5 (*a*) and (*b*) on page 62.

Positioner The deprecated term for a manipulator. See Manipulator for more detail.

Post heat Post-weld heat treatment (PWHT) is any heat treatment given to an arc welded component following welding. In resistance welding it is the provision of a heating cycle following welding and cooling while the component is still in the welding machine.

Powder cutting Oxygen cutting in which a powder, usually finely divided iron for cutting stainless steel, is injected into the oxygen stream to assist cutting by a fluxing action.

Preheat In arc welding the heating of the weld region prior to welding for the purpose of either, (*a*) assisting fusion with thick or highly conductive metals or where there is a difference in thickness between the parts to be joined, or, (*b*) to provide a slower cooling cycle following welding, thereby modifying the final microstructure in the HAZ to avoid hydrogen cracking or hard microstructures. In resistance welding a preliminary current cycle intended to fulfil function (*a*) above.

Preprogrammed power source A power source the output of which, together with such parameters as wire-feed speed, can be set to any one of a number of programs stored within the memory of the power source control system.

Pressure welding Any welding process in which a weld is made by the application of sufficient pressure to cause plastic flow at the faying surfaces which may or may not be heated.

Projection welding Resistance welding in which the localising of force and current to make the weld is obtained by the use of a projection or projections raised on one or more of the faying surfaces or by the natural shape of the component parts of the workpiece (as in cross-wire welding).

Regulator A pressure regulator attached to a cylinder of pressurised gas to reduce and regulate the gas pressure to that required for use.

Resistance welding The generic term for electric welding processes in which the heat for welding is generated by the passage of welding current through a point of locally high electrical resistance created by pressure from electrodes (spot and seam welding), or projections (projection welding), or by pressure applied across the whole area of the members being joined (resistance butt welding).

Reverse polarity A term, used in the USA to indicate electrode positive, but in the UK to indicate electrode negative (see Polarity).

Roller-spot welding Resistance welding in which pressure is applied to the workpiece by a continuously rotating roller on one side and a roller or conducting bar on the other, current being passed intermittently to create a line of spot welds. In step-by-step roller-spot welding the wheels are stopped while the welding current passes.

Root The underside of the first run in a weld or the place this will be in a joint prepared for welding.

Root-face The portion of a fusion face at the root of a joint which is not bevelled or grooved.

Root-run The first run deposited in a multi-run weld.

Run The metal melted or deposited in a single passage of an electrode, torch or blowpipe.

Rutile electrode A metal arc electrode the covering of which contains a high proportion of the mineral rutile.

Seam welding Similar to roller spot welding but with speed and current settings to produce a continuous weld. In step-by-step seam welding the wheels are stopped while the welding current flows to make a series of overlapping spots. In the United States of America, any continuous weld, resistance or fusion.

Self-adjusting arc Arc length control in metal arc welding, in which a small diameter electrode wire is fed continuously at a constant speed, the arc length being held steady by an electrical phenomenon depending on the output characteristic of the power source. Changes in arc length cause changes in welding current which alter the burnoff rate in such a way as to oppose the change and restore the original arc length. See flat characteristic power source.

Semi-automatic welding Welding in which the welding parameters are controlled automatically but manual guidance is necessary.

Series welding Usually refers to resistance spot welding in which two welds are made simultaneously in electrical series, often with the purpose of allowing a simple conducting bar on the under, inaccessible side. Also an arrangement for submerged arc welding where high deposition rate and little penetration is required in which two welding heads feed electrodes into the same molten pool.

Short-circuit transfer See Dip transfer.

Single-shot process A welding process in which a complete joint is made in one operation as opposed to the movement along the joint line of a local heat source. Examples are MIAB, flash and friction welding.

Sintered flux Flux for submerged arc welding which has received a high-temperature treatment sufficient to frit but not fuse the powdered constituents.

Site weld A weld made at the location where the assembly is to be installed.

Slag A fused non-metallic residue produced in some welding processes.

Spatter Globules of metal expelled during welding or cutting.

Spot weld A weld made at a single point between two or more overlapping parts by fusing either through the parts from outside (as in arc spot or laser spot welding) or more usually by a nugget between the parts (as in resistance spot welding).

Spray transfer Metal transfer which takes place in arc welding as a rapidly projected stream of droplets of diameter not greater than that of the consumable electrode from which they are transferred.

Stack cutting The thermal cutting of a stack of plates which are usually clamped together.

Stick electrode A straight covered electrode for MMA welding having the flux removed from one end for insertion in the electrode holder.

Stickout Also known as electrode extension. The length of a continuous consumable electrode projecting beyond the current contact tube. (See I^2R heating.)

Stitch weld In the United Kingdom, a continuous weld made from a succession of overlapping spots or nuggets. Applies to resistance, arc and laser welds. In the United States of America, the term is applied to intermittent welds.

Stop-start defects Defects in a region where welding has been halted temporarily.

Straight polarity A deprecated term for DCEP (in USA) or DCEN (in UK).

Stringer bead A run of weld metal made with little or no weaving motion.

Strip electrodes A thin metal strip used as an electrode in submerged arc surfacing.

Stubbing The freezing of an electrode in the weld.

Stub-end The remains of a stick electrode discarded after the usable length has been deposited.

Stud weld A weld made by a suitable single shot process for joining a metal stud or similar part over its full cross section to a workpiece. Arc, resistance or friction welding can be used.

Submerged arc welding Metal-arc welding with a continuous wire or strip electrode with the arc or arcs submerged under a layer of granular flux some of which is fused in making the weld.

Synergic welding Pulsed-gas metal-arc welding with an electronic power source programmed so that at any wire-feed speed all other parameters, such as current, pulse amplitude and pulse frequency are automatically adjusted instantaneously to provide consistently the most suitable welding conditions. The term is not a synonym for one-knob operation but has been used for non-pulse operation.

Tack weld A weld used to assist assembly or maintain the alignment of edges during welding.

Thermit welding The proprietary name for a process in which union is created between parts by pouring superheated metal between them, the melted metal being generated by an aluminothermic reaction between the oxide of the metal being joined and powdered aluminium. The whole enclosed in a mould to retain the liquid metal and shape the joint. Applied to steel, cast iron and copper. Preferred name aluminothermic welding.

Throat depth The unobstructed distance in a resistance welding machine from the centreline of the electrodes to the nearest point of obstruction. In effect the reach of the machine, and used in this sense for other processes.

Throat thickness (actual) In a fusion weld the perpendicular distance between two lines, each parallel to a line joining the outer toes of the weld, one being a tangent at the weld face and the other being through the furthermost point of fusion penetration. It is the same in both the United States of America and the United Kingdom.

Throat thickness (design) The minimum dimension of throat thickness used for the purposes of design in the United Kingdom only. (See BS 499 for diagrams and further details.)

Thyristor power source A welding power source of the electronic solid-state type using thyristors to control the current.

TIG welding Gas-shielded arc welding using a non-consumable tungsten electrode.

Torch A combined electrode holder and gas nozzle to convey current to the electrode and gas to shield the arc and weld area. Applied to gas tungsten arc TIG and plasma-welding equipment and, in the United States of America, to gas welding equipment, but not in the United Kingdom. The comparable equipment for gas metal arc welding is called a welding gun. Cf Blowpipe.

Transfer efficiency The degree to which alloying elements in a filler metal or electrode are transferred to undiluted weld metal, expressed as a percentage.

Transformer–rectifier A power source comprising a transformer and a rectifier, usually selenium or silicon, which supplies DC for welding.

Transistor power source A power source with associated electronic controls comprising a transformer and a bank of transistors for supplying DC or pulsed current for welding.

Transition piece A pre-prepared insert used for joining two dissimilar metal members of a workpiece, made by a special welding or brazing technique and having ends which match the members to which it is subsequently joined by conventional welding methods.

Ultrasonic welding A pressure-welding process employing mechanical vibrations usually at frequencies above the audible limit superimposed on a static force. Widely used in the electronics and electrical industries.

Undercut An irregular groove caused by welding at a toe of a run in the parent metal or in previously deposited metal.

Upset Parent metal proud of the normal surfaces of the work as a result of forging or pressing. (See Flash, a term also used in drop forging.)

Vertical welding See Positional welding.

Weldability A qualitative term summing up the ease with which a metal may be welded.

Weld bead A single run of weld metal on a surface. A penetration bead is that protruding through the root of a fusion weld made from one side.

Weld bonding Resistance spot welding with adhesive preplaced between the sheets, the main function of the spot welds being to hold the parts while the adhesive is cured.

Welding gun A welding device through which filler wire is fed, e.g. an MIG welding gun. Cf Torch. In resistance spot welding, a pair of portable electrode clamps.

Welding head Originally the device used in automatic welding plant comprising an electrode feed mechanism and means for conveying current to the electrode but now applied to tungsten arc equipment with or without filler wire feed or gas metal arc.

Weld metal All metal melted during the making of a weld and retained in the weld.

Weld pool Preferred term is molten pool and is the pool of liquid metal formed during fusion welding but in electroslag welding this includes the molten slag bath.

Weld toe The edge of the weld face where it meets parent metal or a previous run.

Sources of information

The Welding Institute (TWI), Abington, Cambridge CB1 6AL, UK, and the American Welding Society (AWS), 550 NW Le Jeune Rd, Miami, Florida 33126, USA, are both major publishers of information on welding in the English language. TWI publishes specialised reference lists based on its information database *Weldasearch* which are continually added to and updated. It also publishes books on all aspects of welding and the proceedings of the conferences which it organises and maintains a database on welding consumables. A range of audio and visual aids for welding and non-destructive testing is marketed. Three monthly journals are produced, *Metal Construction* (now retitled *Joining and Materials*), *Welding Abstracts*, with approximately 500 abstracts per month from the world's welding literature, and *Welding International*, providing English translations of selected foreign-language articles published in the major welding journals. Training courses in welding technology and non-destructive testing are organised.

AWS organises conferences, compiles standards and publishes books and other material on welding. The *Welding Journal* incorporating articles of both general and research interest is produced monthly. A publication of major importance is the *Welding Handbook* in five volumes and regularly updated. Other significant publishers of information on welding in the USA are the Welding Research Council (WRC) and the American Society for Metals (ASM). WRC organises research, produces books, monographs, a monthly journal and a *Bulletin* series in which particular welding topics are reviewed. ASM organises conferences on welding, publishing the proceedings, and the *Metals Handbook* devotes one volume to welding, brazing and cutting.

Other English language publications on welding are:

Welding Design and Fabrication – Penton Publications Inc., Cleveland, Ohio, USA.
Welding and Metal Fabrication – IPC Industrial Press Ltd, Guildford, UK.
Welding Review – Fuel and Metallurgical Journals Ltd, Queensway, Redhill, Surrey RH1 1QS, UK.

Most national welding societies publish journals in the language of the country, some of the more important being:

Australian Welding Institute, Eagle House, 118 Alfred Street, Milsons Point, 2061, NSW, Australia: *The Australian Welding Journal.*
Institute Belge de la Soudure, 21 Rue de la Drapiers, Brussels 5, Belgium: *Revue de la Soudure.*
Institut de Soudure, 32 Boulevard de la Chapelle, 75880 Paris, France: *Soudage et techniques connexes.*
Deutscher Verband für Schweisstechnik, 4-Dusseldorf, 172 Aachener Strasse, Postschliessfach 2725, Federal Republic of Germany; *Schweisen und Schneiden*, also *De Practika.*
Japan Welding Society, 1–11 Kanda, Sakuma-cho, Chiyoda-ku, Tokyo, Japan: *Japan Welding Journal.*
Svetstekniska Foreningen, IVA Box 5073, S–102 42 Stockholm 5, Sweden: *Svetsen.*

The national welding societies often provide training courses in various aspects of welding. They can usually give information also on the welding operator schools run by the various manufacturers or other national training bodies.

Bibliography

American Society for Metals, *Metals Handbook*, vol. 6, 'Welding and brazing', 9th edition. Metals Park, Ohio, ASM, 1983, 1152 pp.

American Society for Metals, *Plasma, electron and laser beam technology*, by Y. Arata. Metals Park, Ohio, ASM, 1986, 620 pp.

American Society for Metals, *Worldwide guide to equivalent irons and steels*, edited by Unterweiser, P. M., and Penzenik, M. Metals Park, Ohio, ASM, 1979, 575 pp.

American Society for Metals, *Worldwide guide to equivalent nonferrous metals and alloys*, edited by Unterweiser, P. M., and Penzenik, M. Metals Park, Ohio, ASM, 1980, 625 pp.

American Welding Society, *Brazing Manual*. Miami, Fl, AWS, 1976, 309 pp.

American Welding Society, *Filler metal comparison charts*. Miami, Fl. AWS, 1986, 284 pp.

American Welding Society, *Recommended practices for gas metal arc welding*. Miami, Fl. AWS, 1979, 58 pp.

American Welding Society, *Recommended practices for gas tungsten arc welding*. Miami, Fl. AWS, 1980, 38 pp.

American Welding Society, *Recommended practices for electrogas welding*. Miami, Fl. AWS, 1981, 46 pp.

American Welding Society, *Weldability of steels*, 4th edition, by Stout, R. D. Miami, Fl. AWS, 1987, 450 pp.

American Welding Society, *Welding Handbook*, 8th edition, vol 1, 'Welding Technology' (first volume of new edition).

Welding Handbook, 7th edition:

vol. 2, 'Welding processes, Arc, gas, cutting, brazing and soldering.'

vol. 3, 'Welding processes, Resistance, solid state, and others.'

vol. 4. 'Metals and their weldability.'

vol. 5. 'Engineering costs, quality and safety.'

American Welding Society, *Welding inspection*. Miami, Fl. AWS, 1980, 222 pp.

Castro, R. J., and Cadenet, J. J. de, *Welding metallurgy of stainless and heat-resisting steels*, trans. from French. London, CUP, 1974, 190 pp.

Crossland, B., *Explosive welding of metals and its application*. Oxford, Clarendon Press, 1982, 250 pp.

Dawson, R. J. C., *Fusion welding and brazing of copper and copper alloys*. London, Newnes-Butterworths, 1973, 139 pp.

Hammond, R., *Glossary of welding terms* (A five-language dictionary and glossary). London, Frederick Muller Ltd, 1974, 227 pp.

Hicks, J. G., *Welded joint design*, London, Crosby Lockwood Staples, 1979, 82 pp.

Houldcroft, P. T., *Welding process technology*. Cambridge, CUP, 1977, 313 pp.

Lancaster, J. F. (editor), *Physics of welding*. Second edition, Oxford, Pergamon Press for International Institute of Welding, 1986, 340 pp.

Lancaster, J. F., *Metallurgy of welding*. Fourth edition, London, Allen and Unwin, 1987, 361 pp.

Malin, V., *Monograph on narrow-gap welding technology*. Bulletin No. 323, Welding Research Council, New York, 1987.

Paton, B. E. (editor), *Electroslag welding and surfacing*, vols 1 and 2, English trans. Moscow, MIR Publishers, 1983, 256 and 264 pp.

Weymueller, C. R., *Know your welding NDT*. Cleveland, Ohio, Penton IPC, 1981, 153 pp.

Welding Design and Fabrication. 1987 ninth annual buyer's guide to welding and fabricating products. *Welding Design and Fabrication* vol. 60, No. 1, January 1987.

Welding Institute, *Control of distortion in welded fabrications*. Abington, TWI, 1967, 70 pp.

Welding Institute, *Health and safety in welding and allied processes*. Abington, TWI, 1983, 207 pp.

Welding Institute, *Welding cast irons*. Abington, TWI, 1986, 40 pp.

Welding Institute, *Facts about fume* (N. Jenkins, editor). Abington, TWI, 1986, 40 pp.

Welding Institute, *Laser welding, cutting and surface treatment*. Abington, TWI, 1984, 58 pp.

Welding Institute, *Welding dissimilar metals*. Abington, TWI, 1986, 69 pp.

Index